Y0-BGG-626

Principles of

APPLIED RESERVOIR

SIMULATION

Principles of
APPLIED RESERVOIR
SIMULATION

John R. Fanchi

Gulf Publishing Company
Houston, Texas

Principles of
APPLIED RESERVOIR SIMULATION

Copyright © 1997 by Gulf Publishing Company, Houston, Texas. All rights reserved. This book, or parts thereof, may not be reproduced in any form without express written permission of the publisher. The software—BOAST 4D—is licensed for the exclusive use of the original purchaser on one computer only.

Gulf Publishing Company
Book Division
P.O. Box 2608 ☐ Houston, Texas 77252-2608

10 9 8 7 6 5 4 3 2 1

Library of Congress Cataloging-in-Publication Data

Fanchi, John R.
    Principles of applied reservoir simulation / John R. Fanchi.
      p. cm.
    Includes bibliographical references and index.
    ISBN 0-88415-117-4 (alk. paper)
    1. Oil fields—Computer simulation. 2. Petroleum—
Geology—Mathematical models. I. Title.
    TN870.53.F36   1997
    622′.3382′0113—DC21   97-39963
                                                  CIP

Printed in the United States of America

Printed on Acid-Free Paper (∞)

To my parents,
John A. and Shirley M. Fanchi

Contents

Part II
Case Study

Part IV
BOAST4D Technical Supplement

About the Author

John R. Fanchi, Ph.D.

John Fanchi has B.S., M.S., and Ph.D. degrees in physics from the universities of Denver, Mississippi, and Houston, respectively. His oil industry responsibilities have revolved around reservoir modeling, both in the areas of simulator development and reservoir management applications. In addition to being the principal author of the U.S. Department of Energy simulator BOAST and its successor BOAST II, he has performed development work on compositional, electromagnetic heating, chemical flood, and geothermal simulators. His reservoir management experience includes project leadership and significant participation in studies of oil, gas and condensate fields in the North Sea, offshore Sakhalin Island, the Persian Gulf, the Gulf of Mexico, Alaska, and in many parts of the mainland U.S. These studies include preparing models of primary, secondary, and enhanced recovery applications.

Dr. Fanchi has designed and taught courses in applied reservoir simulation, waterflooding, reservoir engineering, natural gas engineering, black oil simulation, compositional simulation, and history matching. His publication credits include *Math Refresher for Scientists and Engineers, Parametrized Relativistic Quantum Theory,* and numerous technical publications.

Preface

Principles of Applied Reservoir Simulation is a vehicle for widely disseminating reservoir simulation technology. It is not a mathematical treatise about reservoir simulation, nor is it a compendium of case histories. Both of these topics are covered in several other readily available sources. Instead, *Principles of Applied Reservoir Simulation* is a practical guide to reservoir simulation that introduces the novice to the process of reservoir modeling and includes a fully functioning reservoir simulator for the reader's personal use.

Part I explains the concepts and terminology of reservoir simulation. The selection of topics and references is based on what I have found to be most useful over the past two decades as both a developer and user of reservoir simulators. I have provided advice gleaned from model studies of a variety of oil, gas, and condensate fields.

Participation is one of the best ways to learn a subject. The exercises in Part I let you apply the principles that are discussed in each chapter. As a means of integrating the material, Part II applies the principles of reservoir simulation to the study of a particular case. By the time you have completed the case study, you will have participated in each technical phase of a typical model study.

Parts III and IV are the User's Manual and Technical Supplement, respectively for the three-dimensional, three-phase black oil simulator BOAST4D that accompanies the text. BOAST4D is a streamlined and upgraded version of BOAST II, a public domain black oil simulator developed for the U.S. Department of Energy in the 1980's. As principal author of BOAST II, I have added several features and made corrections to create BOAST4D. For example, you can now use BOAST4D to model horizontal wells and perform reservoir geophysical calculations. The latter calculations are applicable to an emerging technology: 4D seismic monitoring of fluid flow. The inclusion of reservoir geophysical calculations is the motivation for appending "4D" to the program name. In addition, BOAST4D includes code changes to improve computational performance, to solve material balance problems, and to reduce material balance error.

BOAST4D was designed to run on DOS-based personal computers with 486 or better math co-processors. The simulator included with this book is well-suited for learning how to use a reservoir simulator, for developing an understanding of reservoir management concepts, and for solving many types of reservoir engineering problems. It is an inexpensive tool for performing studies that require more sophistication than is provided by analytical solutions, yet do not require the use of full-featured commercial simulators. Several example data sets are provided on disk to help you apply the simulator to a wide range of practical problems.

The text and software are suitable for use in a variety of settings, e.g., in an undergraduate course for petroleum engineers, earth scientists such as geologists and geophysicists, or hydrologists; in a graduate course for modelers; and in continuing education courses. An Instructor's Guide for qualified instructors is available from the publisher.

I developed much of the material in this book as course notes for a continuing education course I taught in Houston. I would like to thank Bob Hubbell and the University of Houston for sponsoring this course and Tim Calk of Gulf Publishing Company for shepherding the manuscript through the publication process. I am grateful to my industrial and academic employers, both past and present, for the opportunity to work on a wide variety of problems. I would also like to acknowledge the contributions of Ken Harpole, Stan Bujnowski, Jane Kennedy, Dwight Dauben, and Herb Carroll for their work on earlier versions of BOAST. Special thanks go to my wife, Kathy Fanchi, for her moral support and many hours at the computer creating the graphics and refining the presentation of this material.

Any written comments or suggestions for improving the material are welcome.

John R. Fanchi, Ph.D.
Houston, Texas

Part I

Principles

Chapter 1

Introduction to Reservoir Management

Reservoir modeling exists within the context of the reservoir management function. Although not universally adopted, reservoir management is often defined as the allocation of resources to optimize hydrocarbon recovery from a reservoir while minimizing capital investments and operating expenses [Wiggins and Startzman, 1990; Satter, et al., 1994; Al-Hussainy and Humphreys, 1996; Thakur, 1996]. These two outcomes – optimizing recovery and minimizing cost – often conflict with each other. Hydrocarbon recovery could be maximized if cost was not an issue, while costs could be minimized if the field operator had no interest in or obligation to prudently manage a finite resource. *The primary objective in a reservoir management study is to determine the optimum conditions needed to maximize the economic recovery of hydrocarbons from a prudently operated field.* Reservoir modeling is the most sophisticated methodology available for achieving the primary reservoir management objective.

There are many reasons to perform a model study. Perhaps the most important, from a commercial perspective, is the ability to generate cash flow predictions. Simulation provides a production profile for preparing economic forecasts. The combination of production profile and price forecast gives an estimate of future cash flow. Other reasons for performing a simulation study from a reservoir management perspective are listed in Table 1-1. Several of the items are discussed in greater detail in later chapters.

Table 1-1

Why Simulate?

Corporate Impact
♦ Cash Flow Prediction
◇ Need Economic Forecast of Hydrocarbon Price
Reservoir Management
♦ Coordinate Reservoir Management Activities
♦ Evaluate Project Performance
◇ Interpret/Understand Reservoir Behavior
♦ Model Sensitivity to Estimated Data
◇ Determine Need for Additional Data
♦ Estimate Project Life
♦ Predict Recovery vs Time
♦ Compare Different Recovery Processes
♦ Plan Development or Operational Changes
♦ Select and Optimize Project Design
◇ Maximize Economic Recovery

1.1 Consensus Modeling

Reservoir modeling is the application of a computer simulation system to the description of fluid flow in a reservoir [for example, see Peaceman, 1977; Aziz and Settari, 1979; Mattax and Dalton, 1990]. The computer simulation system is usually just one or more computer programs. To minimize confusion in this text, the computer simulation system is called the reservoir simulator, and the input data set is called the reservoir model.

Many different disciplines contribute to the preparation of the input data set. The information is integrated during the reservoir modeling process, and the concept of the reservoir is quantified in the reservoir simulator. Figure 1-1 illustrates the contributions different disciplines make to reservoir modeling.

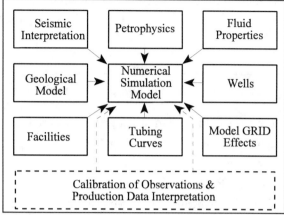

Figure 1-1. Disciplinary contributions to reservoir modeling (after H.H. Haldorsen and E. Damsleth, ©1993; reprinted by permission of the American Association of Petroleum Geologists).

The simulator is the point of contact between disciplines. It serves as a filter that selects from among all of the proposed descriptions of the reservoir. The simulator is not influenced by hand-waving arguments or presentation style. It provides an objective appraisal of each hypothesis, and constrains the power of personal influence described by Millheim [1997]. As a filter of hypotheses, the reservoir modeler is often the first to know when a proposed hypothesis about the reservoir is inadequate.

One of the most important tasks of the modeler is to achieve consensus in support of a reservoir representation. This task is made more complex when available field performance data can be matched by more than one reservoir model. The non-uniqueness of the model is discussed in greater detail through-out the text. It means that there is more than one way to perceive and represent available data. The modeler must sort through the various reservoir represen-tations and seek consensus among all stakeholders. This is often done by rejecting one or more proposed representations. As a consequence, the human element is a factor in the process, particularly when the data do not clearly support the selection of a single reservoir representation from a set of competing representations. The dual criteria of reasonableness and Ockham's Razor

[Chapter 9.3; Jefferys and Berger, 1992] are essential to this process, as is an understanding of how individuals can most effectively contribute to the modeling effort.

1.2 Management of Simulation Studies

Ideally, specialists from different disciplines will work together as a team to develop a meaningful reservoir model. Team development proceeds in well known stages [Sears, 1994]:
- ◆ Introductions: Getting to know each other
- ◆ "Storming": Team members disagree over how to proceed
    - ◇ Members can lose sight of goals
- ◆ "Norming": Members set standards for team productivity
- ◆ "Performing": Team members understand
    - ◇ what each member can contribute
    - ◇ how the team works best

Proper management recognizes these stages and allows time for the team building process to mature.

Modern simulation studies of major fields are performed by teams that function as project teams in a matrix management organization. Matrix management is synonymous here with Project Management and has two distinct characteristics:
- ◆ "Cross-functional organization with members from different work areas who take on a project." [Staff-JPT, 1994]
- ◆ "One employee is accountable to two or more superiors, which can cause difficulties for managers and employees." [Staff-JPT, 1994]

To alleviate potential problems, the project team should be constituted such that
- ◆ Each member of the team is assigned a different task.
- ◆ All members work toward the same goal.

Team members should have unique roles to avoid redundant functions. If the responsibilities of two or more members of the team overlap considerably, confusion may ensue with regard to areas of responsibility and, by implication, of accountability. Each team member must be the key decision maker in a particular discipline, otherwise disputes may not get resolved in the time available for completing a study. Teams should not be allowed to flounder in an egalitarian utopia that does not work.

Effective teams may strive for consensus, but the pressure of meeting deadlines will require one team member to serve as team leader. Deadlines cannot be met if a team cannot agree, and there are many areas where decisions may have to be made that will not be by consensus. For this reason, teams should have a team leader with the following characteristics:

♦ Significant technical skills

♦ Broad experience

Team leaders should have technical and monetary authority over the project. If they are perceived as being without authority, they will be unable to fulfill their function. On the other hand, team leaders must avoid authoritarian control or they will weaken the team and wind up with a group.

According to Maddox [1988], teams and groups differ in the way they behave. Group behavior exhibits the following characteristics:

♦ "Members think they are grouped together for administrative purposes only. Individuals work independently, sometimes at cross purposes."

♦ "Members tend to focus on themselves because they are not sufficiently involved in planning the unit's objectives. They approach their job simply as hired hands."

By contrast, the characteristics of team behavior are the following:

♦ "Members recognize their interdependence and understand both personal and team goals are best accomplished with mutual support. Time is not wasted struggling over territory or seeking personal gain at the expense of others."

♦ "Members feel a sense of ownership for their jobs and unit because they are committed to goals they helped to establish."

Similar observations were made by Haldorsen and Damsleth [1993]:

♦ "Members of a team should necessarily understand each other, respect each other, act as a devil's advocate to each other, and keep each other informed."

Haldorsen and Damsleth [1993] argue that the focus of each team member should be:

♦ Innovation and creation of value through the team approach

♦ Customer orientation with focus on "my output is your input"

McIntosh, et al. [1991] support the notion that each team member should fulfill a functional role, for example, geoscientist, engineer, etc. A corollary is that team members can understand their roles because the roles have been clearly defined.

Proper management can improve the likelihood that a team will function as it should. A sense of ownership or "buy-in" can be fostered if team members participate in planning and decision making. Team member views should influence the work scope and schedule of activity. Many problems can be avoided if realistic expectations are built into project schedules at the beginning, and then adhered to throughout the project. Expanding work scope without altering resource allocation or deadlines can be demoralizing and undermine the team concept.

Finally, one important caution should be borne in mind when performing studies using teams: "Fewer ideas are generated by groups than by individuals working alone – a conclusion supported by empirical evidence from psychology. [Norton, 1994]" Smoothly functioning teams may be productive, but they may also exhibit a lack of creativity. Brainstorming sessions or cross-fertilization of ideas by occasionally rearranging team assignments may need to be a part of team management.

1.3 Outline of the Text

The remainder of the text is organized as follows. The material in Part I explains the concepts and terminology of reservoir simulation. A typical exercise in Part I asks you to find and change data records in a specified example data file. These records of data must be modified based on an understanding of the reservoir problem and a familiarity with the accompanying computer program BOAST4D. BOAST4D is a three-dimensional, three-phase reservoir simulator. These terms will be discussed in detail in subsequent chapters.

The exercises in Part I use different sections of the user's manual presented in Part III. If you work all the exercises, you will be familiar with the user's manual and BOAST4D by the time the exercises are completed. Much of the experience gained by running BOAST4D is applicable in principle to other simulators.

Successful completion of the exercises in Part I will prepare you for the case study presented in Part II. The case study is designed to integrate the material discussed in Part I. By the time Part II is completed, you will have participated in each technical phase of a typical model study.

Parts III and IV are the User's Manual and Technical Supplement, respectively, for BOAST4D. Supplemental information in Part IV provides more detailed descriptions of the algorithms coded in BOAST4D.

Exercises

Exercise 1.1 BOAST4D Directory: A three-dimensional, three-phase reservoir simulator (BOAST4D) is included as a disk with this book. The BOAST4D user's manual is presented in Part III, and a technical supplement is provided in Part IV. Prepare a directory on your hard drive for running BOAST4D using the procedure outlined in Chapter 18.

Exercise 1.2 BOAST4D Example Data Sets: Several example data sets are provided on the BOAST4D disk. They are briefly described in Chapter 22, which also presents a complete example data set. Copy all files with the ".DAT" extension from your disk to the \BOAST4D directory on your hard drive. This includes data sets of the form EXAM*.DAT and CS-*.DAT. The latter files are case study data files and are used extensively in Part II. Unless stated otherwise, all exercises assume BOAST4D and its data sets reside in the \BOAST4D directory.

Exercise 1.3 The simulator BOAST4D can be run from a disk drive or a hard drive. It is faster to load and run BOAST4D from the hard drive. To demonstrate the difference, copy file EXAM7.DAT to BTEMP.DAT and then run BOAST4D from both the disk drive and the hard drive.

References

Al-Hussainy, R. and N. Humphreys (1996): "Reservoir Management: Principles and Practices," *Journal of Petroleum Technology*, pp. 1129-1135.

Aziz, K. and A. Settari (1979): **Petroleum Reservoir Simulation**, New York: Elsevier.

Haldorsen, H.H. and E. Damselth (1993): "Challenges in Reservoir Characterization," *American Association of Petroleum Geologists Bulletin*, Volume 77, No. 4, pp. 541-551.

Jefferys, W.H. and J.O. Berger (1992): "Ockham's Razor and Bayesian Analysis," *American Scientist*, Volume 80, pp. 64-72.

Maddox, R.B. (1988): **Team Building: An Exercise in Leadership**, Crisp Publications, Inc.

Mattax, C.C. and R.L. Dalton (1990): **Reservoir Simulation**, SPE Monograph #13, Richardson, TX: Society of Petroleum Engineers.

McIntosh, I., H. Salzew, and C. Christensen (1991): "The Challenge of Teamwork," Paper CIM/AOSTRA 91-19, *Proceedings of CIM/AOSTRA 1991 Technical Conference*, Banff, Canada.

Millheim, K.M. (1997): "Fields of Vision," *Journal of Petroleum Technology*, p. 684.

Norton, R. (Nov. 1994): "Economics for Managers," *Fortune*, pg. 3.

Satter, A., J.E. Varnon, and M.T. Hoang (1994): "Integrated Reservoir Management," *Journal of Petroleum Technology*, pp. 1057-1064.

Sears, M. (June 1994): organizational development specialist at Bell Atlantic, quoted in *Journal of Petroleum Technology*, p. 505.

Staff-JPT (Aug. 1994): "New Management Structures: Flat and Lean, Not Mean," *Journal of Petroleum Technology*, pp. 647-648.

Thakur, G.C. (1996): "What Is Reservoir Management?" *Journal of Petroleum Technology*, pp. 520-525.

Wiggins, M.L. and R.A. Startzman (1990): "An Approach to Reservoir Management," Paper SPE 20747, *Proceedings of 65th Annual Society of Petroleum Engineers Fall Meeting*, New Orleans, LA.

Chapter 2

Overview of the Modeling Process

The process of applying a reservoir flow simulator to the study of a physical system is outlined here. The best technology for making reservoir performance predictions today is to model fluid flow in porous media using computer programs known as simulators.

2.1 Basic Reservoir Analysis

Reservoir characterization and reservoir engineering evaluations are usually performed as a part of standard business practice independent of a reservoir simulation study. The tasks associated with basic reservoir analyses [for example, see Mian, 1992; Tearpock and Bischke, 1991] provide information that is needed to prepare input data for a simulation study. They also provide an initial concept of the reservoir which can be used to design a model study. The modeling team needs to be aware of early studies and should relate model performance to previous studies whenever possible.

2.2 Prerequisites

Several prerequisites should be satisfied before a model study is undertaken [Coats, 1969]. The most important, from a business perspective, is the existence of a problem of economic importance. At the very least, the objectives of a model study should yield a solution to the economically important problem.

Once the objectives of a study are specified, the modeler should gather all available data and reports relating to the field. The term "modeler" is used in the remainder of the text as a synonym for "modeling team" unless an explicit distinction must be made. If necessary data is not available, the modeler should determine if the data can be obtained, either by analogy with other reservoirs or by correlation. Values for all model input data must be obtained because the simulator will not run without a complete set of data. In some cases, simplifying assumptions about the reservoir may have to be made because there is not enough data available to quantitatively represent the system in greater detail.

In addition to clearly defined objectives, another prerequisite that must be satisfied before committing to a simulation study is to determine that the objectives of the study cannot be achieved using simpler techniques. If less expensive techniques, such as decline curve analysis or the Buckley-Leverett waterflood displacement algorithm [Collins, 1961; Craig, 1971; and Dake, 1978], do not provide adequate results, then more sophisticated and costly methods are justified.

2.3 Computer Modeling

A comprehensive reservoir management model can be thought of as four interacting models: the reservoir model, the well model, the wellbore model, and

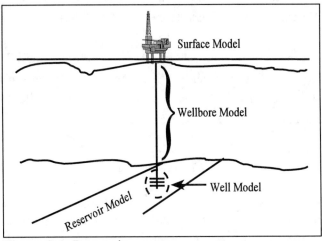

Figure 2-1. Reservoir management system.

the surface model. The spatial relationship between these models is illustrated in Figure 2-1. The reservoir model represents fluid flow within the reservoir. The reservoir is modeled by subdividing the reservoir volume into an array, or grid, of smaller volume elements (Figure 2-2). Many names are used to denote the individual volume elements, for example, grid block, cell or node. The set of all volume elements is known by such names as grid or mesh.

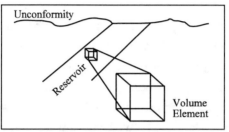

Figure 2-2. Subdivide reservoir.

Every practical reservoir simulator includes both a reservoir model and a well model. The well model is a term in the fluid flow equations that represents the extraction of fluids from the reservoir or the injection of fluids into the reservoir. Full featured commercial simulators also include a wellbore model and a surface facility model. The wellbore model represents flow from the sandface to the surface. The surface model represents constraints associated with surface facilities, such as platform and separator limitations.

The mathematical algorithms associated with each model depend on physical conservation laws and empirical relationships. Computer simulators are based on conservation of mass, momentum, and energy. The most widely used simulators assume the reservoir is isothermal, that is, constant temperature. If we are modeling a reservoir where thermal effects matter, such as a secondary recovery process where heat has been injected in some form, then we need to use a simulator that accounts for temperature variation and associated thermodynamic effects. The set of algorithms is sufficiently complex that high speed computers are the only practical means of solving the mathematics associated with a reservoir simulation study. These topics are discussed in more detail in later chapters.

2.4 Major Elements of a Reservoir Simulation Study

The essential elements of a simulation study include matching field history; making predictions, including a forecast based on the existing operating strategy; and evaluating alternative operating scenarios [Mattax and Dalton, 1990; Thomas, 1982]. During the history match, the modeler will verify and refine the reservoir description. Starting with an initial reservoir description, the model is used to match and predict reservoir performance. If necessary, the modeler will modify the reservoir description until an acceptable match is obtained. The history matching phase of the study is an iterative process that makes it possible to integrate reservoir geoscience and engineering data.

The history matching process may be considered an inverse problem because an answer already exists. We know how the reservoir performed; we want to understand why. Our task is to find the set of reservoir parameters that minimizes the difference between the model performance and the historical performance of the field. This is a non-unique problem since there is usually more than one way to match the available data.

Once a match of historical data is available, the next step is to make a base case prediction, which is essentially just a continuation of existing operating practice. The base case prediction gives a baseline for comparison with other reservoir management strategies.

Model users should be aware of the validity of model predictions. One way to get an idea of the accuracy of predictions is to measure the success of forecasts made in the past. Lynch [1996] looked at the evolution of the United States Department of Energy price forecast over a period of several years for both oil and gas. The quality of price forecasts is illustrated in Figure 2-3. Forecasts that were made in years 1981, 1984, 1987, and 1991 are compared to the actual prices. Even though price forecast is essential to a commercial enterprise, it is clear from Lynch's study that there is considerable uncertainty associated with the price forecast. Forecasts do not account for discontinuities in historical patterns that arise from unexpected effects. This is as true in the physical world as it is in the social [Oreskes, et al., 1994]. In addition to uncertainty in economic parameters, there is uncertainty in the forecasted production performance of a field. Simulators do not eliminate uncertainty; they give us the ability to assess and better manage the risk associated with the prediction of production performance.

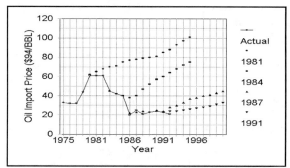

Figure 2-3. Price forecasting.

A valuable but intangible benefit of the process associated with reservoir simulation is the help it provides in managing the reservoir. One of the critical tasks of reservoir management is the acquisition and maintenance of an up-to-date data base. A simulation study can help coordinate activities as a modeling team gathers the resources it needs to determine the optimum plan for operating a field. Collecting input data for a model is a good way to ensure that every important technical variable is considered as data is collected from the many disciplines that contribute to reservoir management. If model performance is especially sensitive to a particular parameter, then a plan should be made to determine that parameter more accurately, for example, from either laboratory or appropriate field tests.

Exercises

Exercise 2.1 Original Volume In Place: Data file EXAM1.DAT is a material balance model of an undersaturated oil reservoir undergoing pressure depletion. Run EXAM1.DAT and find the volume of oil and gas originally in place. Remember to copy EXAM1.DAT to BTEMP.DAT before executing BOAST4D. The fluid volumes of interest are reported in output file BTEMP.OUT. They can be found by using an ASCII editor to search through BTEMP.OUT.

Exercise 2.2 Gas Reservoir Material Balance: Suppose a gas reservoir has the following production history:

G_p (Bscf)	P (psia)	Z	P/Z (psia)
0.015	1946	0.813	2393
0.123	1934	0.813	2378
0.312	1913	0.814	2350
0.652	1873	0.815	2297
1.382	1793	0.819	2190
2.21	1702	0.814	2091
2.973	1617	0.828	1953
3.355	1576	0.83	1898
4.092	1490	0.835	1783
4.447	1454	0.838	1734
4.822	1413	0.841	1679

where G_p is cumulative gas production, P is pressure, and Z is gas compressibility factor. Draw a straight line through a plot of G_p vs P/Z to find original gas in place (OGIP). OGIP corresponds to $P/Z = 0$. These results were obtained from data file EXAM8.DAT. Verify that the OGIP for the model is about 15.9 Bscf by running EXAM8.DAT and finding the OGIP in BTEMP.OUT. How much oil and water are originally in place?

References

Coats, K.H. (1969): "Use and Misuse of Reservoir Simulation Models," *Journal of Petroleum Technology*, pp. 183-190.

Collins, R.E. (1961): **Flow of Fluids Through Porous Materials**, Tulsa, OK: PennWell Publishing.

Craig, F.F. (1971): **The Reservoir Engineering Aspects of Waterflooding**, SPE Monograph Series, Richardson, TX: Society of Petroleum Engineers.

Dake, L.P. (1978): **Fundamentals of Reservoir Engineering**, Amsterdam: Elsevier.

Lynch, M.C. (1996): "The Mirage of Higher Petroleum Prices," *Journal of Petroleum Technology*, pp. 169-170.

Mattax, C.C. and R.L. Dalton (1990): **Reservoir Simulation**, SPE Monograph #13, Richardson, TX: Society of Petroleum Engineers.

Mian, M.A. (1992): **Petroleum Engineering Handbook for the Practicing Engineer**, Volumes I and II, Tulsa, OK: PennWell Publishing.

Oreskes, N., K. Shrader-Frechette, and K. Belitz (4 Feb. 1994): "Verification, Validation, and Confirmation of Numerical Models in the Earth Sciences", *Science*, pp. 641-646.

Tearpock, D.J. and R.E. Bischke (1991): **Applied Subsurface Geological Mapping**, Englewood Cliffs, NJ: Prentice Hall.

Thomas, G.W. (1982): **Principles of Hydrocarbon Reservoir Simulation**, Boston: International Human Resources Development Corporation.

Chapter 3

Conceptual Reservoir Scales

One of the most important goals of modeling is to reduce the risk associated with making decisions in an environment where knowledge is limited. The range of applicability of acquired data and the integration of scale-dependent data into a cohesive reservoir concept are discussed below.

3.1 Reservoir Sampling and Scales

A sense of just how well we understand the reservoir can be obtained by considering the fraction of reservoir area sampled by different techniques. As an example, suppose we want to find the size of the area sampled by a wellbore that has a six inch radius. If we assume the area is circular, we can calculate the area as πr^2 where r is the sampled radius. The resulting sampled area is less than a square foot. To determine the fraction of area sampled, we normalize the sampled area with respect to the drainage area of a well, say a very modest five acres. What fraction of the area is directly sampled by the wellbore? The drainage area is 218,000 square feet. The fraction of the area sampled by the well is three to four parts in a million. This is a tiny fraction of the area of interest.

A well log signal will expand the area that is being sampled. Suppose a well log can penetrate the formation up to five feet from the wellbore, which is a reasonably generous assumption. The fraction of area that has been sampled is now approximately four parts in 10 thousand. The sample size in a drainage area of five acres, which is a small drainage area, is still a fraction of a percent.

Core and well log information gives us a very limited view of the reservoir. A seismic section expands the fraction of area sampled, but the interpretation of seismic data is less precise. Seismic data is often viewed as "soft data" because of its dependence on interpretation. The reliability of seismic interpretation can be improved when correlated with "hard data" such as core and well log measurements.

The range of applicability of measured data depends on the sampling technique. Did we take some core out of the ground, measure an electrical response from a well log, or detect acoustical energy? The ranges are illustrated in Figure 3-1. Fayers and Hewett [1992] point out that scale definitions are not universally accepted, but do illustrate the relative scale associated with reservoir property measurements. Scale sizes range from the very big to the microscopic. To recognize variations in the range of data applicability, four conceptual scales have been defined (Figure 3-2) and will be adopted for use in the following discussion.

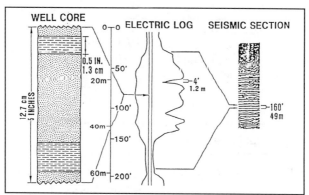

Figure 3-1. Range of data sampling techniques (after Richardson, et al., 1987a; reprinted by permission of the Society of Petroleum Engineers).

The Giga Scale includes information associated with geophysical techniques, such as reservoir architecture. Theories of regional characterization, such as plate tectonics, provide an intellectual framework within which Giga Scale measurement techniques, like seismic and satellite data, can be interpreted. The Mega Scale is the scale of reservoir characterization and includes well logging, well testing and 3D seismic analysis. The Macro Scale focuses on data sampling at the level of core analysis and fluid property analysis. The Micro Scale includes pore scale data obtained from techniques such as thin section analysis and measurements of grain size distribution. Each of these scales contributes to the final reservoir model.

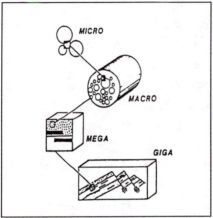

Figure 3-2. Reservoir scales (after Haldorsen and Lake, 1989; reprinted by permission of the Society of Petroleum Engineers).

3.2 Integrating Scales – the Flow Unit

All of the information collected at various scales must be integrated into a single, comprehensive, and consistent representation of the reservoir. The integration of data obtained at different scales is a difficult issue that is often referred to as the "scale-up" problem [for example, see Oreskes, et al., 1994]. Attempts to relate data from two different scales can be difficult. For example, permeability is often obtained from both pressure transient testing and routine core analysis. The respective permeabilities, however, may appear to be uncorrelated because they represent two different measurement scales. An important task of the scale-up problem is to develop a detailed understanding of how measured parameters vary with scale. The focus on detail in one or more aspects of the reservoir modeling process can obscure the fundamental reservoir concept in a model study. One way to integrate available data within the context of a "big picture" is to apply the flow unit concept.

A flow unit is defined as "a volume of rock subdivided according to geological and petrophysical properties that influence the flow of fluids through it" [Ebanks, 1987]. Typical geologic and petrophysical properties are shown in Table 3-1. A classic application of the flow unit concept is presented in a paper by Slatt and Hopkins [1990].

Table 3-1
Properties Typically Needed to Define a Flow Unit

Geologic	Petrophysical
Texture	Porosity
Mineralogy	Permeability
Sedimentary Structure	Compressibility
Bedding Contacts	Fluid Saturations
Permeability Barriers	

A reservoir is modeled by subdividing its volume into an array of representative elementary volumes (REV). The REV concept is not the same as the flow unit concept. A flow unit is a contiguous part of the reservoir that has similar flow properties as characterized by geological and petrophysical data. Flow units usually contain one or more REVs. By contrast, the REV is the volume element that is large enough to provide statistically significant average values of parameters describing flow in the contained volume, but small enough to provide a meaningful numerical approximation of the fundamental flow equations [for example, see Bear, 1972]. As noted by Fayers and Hewett [1992], "It is somewhat an act of faith that reservoirs can be described by relatively few REV types at each scale with stationary average properties."

The flow unit concept is an effective means of managing the growing base of data being provided by geoscientists. Increasing refinement in geoscientific analysis gives modelers more detail than they can use. Even today, with 100,000 grid block flow models, modelers cannot use all of the information that is provided by computer-based geologic models which may be based on over a million grid points. It is still necessary to coarsen detailed geologic models into representative flow units.

An understanding of the big picture, even as a simple sketch, is a valuable resource for validating the ideas being quantified in a model. Richardson, et al. [1987b] sketched several common types of reservoir models: a deep-water fan; a sand-rich delta; a deltaic channel contrasted with a deltaic bar, etc. Their sketches illustrate what the reservoir might look like for a specified set of assumptions. A sketch such as Figure 3-3 is a good tool for confirming that people from different disciplines share the same concept of a reservoir; it is a simple visual aid that enhances communication. In many cases, especially the case of relatively small fields, the best picture of the reservoir may only be a qualitative picture. When a more detailed study begins, the qualitative picture can be upgraded by quantifying parameters such as gross thickness in the context of the conceptual sketch of the reservoir.

Figure 3-3. Mississippi Delta.

Confidence in model performance is acquired by using the model to match historical field performance. History matching and model validation will be discussed in greater detail later. As a rule, models should be updated and refined as additional field data is obtained.

Exercises

Exercise 3.1 Multiply the volume of the reservoir in EXAM1.DAT by 0.1, 10 and 100. This can be done by altering the grid block size (see Chapter 19.1.1). Verify the volume changes by checking initial volumes in BTEMP.OUT. How does the change in volume affect the pressure performance of the model as a function of time?

Exercise 3.2 Repeat Exercise 3.1, but make the volume changes by modifying the grid dimensions using the modification option presented in Chapter 19.1.2.

References

Bear, J. (1972): **Dynamics of Fluids in Porous Media**, New York: Elsevier.

Ebanks, W.J., Jr. (1987): "Flow Unit Concept – Integrated Approach to Reservoir Description for Engineering Projects," paper presented at the AAPG Annual Meeting, Los Angeles.

Fayers, F.J. and T.A. Hewett (1992): "A Review of Current Trends in Petroleum Reservoir Description and Assessing the Impacts on Oil Recovery," *Proceedings of Ninth International Conference on Computational Methods in Water Resources*, June 9-11.

Haldorsen, H.H. and L.W. Lake (1989): "A New Approach to Shale Management in Field-Scale Models," **Reservoir Characterization-2**, SPE Reprint Series #27, Richardson, TX: Society of Petroleum Engineers.

Oreskes, N., K. Shrader-Frechette, and K. Belitz (4 Feb. 1994): "Verification, Validation, and Confirmation of Numerical Models in the Earth Sciences", *Science*, pp. 641-646.

Richardson, J.G., J.B. Sangree, and R.M. Sneider (1987a): "Applications of Geophysics to Geologic Models and to Reservoir Descriptions," *Journal of Petroleum Technology*, pp. 753-755.

Richardson, J.G., J.B. Sangree, and R.M. Sneider (1987b): "Introduction to Geologic Models," *Journal of Petroleum Technology* (first of series), pp. 401-403.

Slatt, R.M. and G.L. Hopkins (1990): "Scaling Geologic Reservoir Description to Engineering Needs," *Journal of Petroleum Technology*, pp. 202-210.

Chapter 4

Reservoir Structure

The physical size and shape of the reservoir may be inferred from several methods that serve as sources of information for defining the large scale structure of the reservoir. These information sources are briefly reviewed below.

4.1 Giga Scale

Seismic measurements discussed in the literature by authors such as Ausburn, et al. [1978], McQuillin [1984], and Sheriff [1989] provide much of the Giga Scale information that can be directly used to characterize a reservoir. Historically, seismic analyses have been of interest primarily as a means of establishing the structural size of the reservoir. People did not believe that seismic data could resolve sufficient detail to provide information beyond overall reservoir structure. But that view has changed with the emergence of 4D seismic monitoring and reservoir geophysics [for example, see Richardson, 1989; Ruijtenberg, et al., 1990; Anderson, 1995; He, et al., 1996; Johnston, 1997]. It is therefore worthwhile to introduce some basic geophysical concepts within the context of the reservoir management function.

Seismic waves are vibrations that propagate from a source, such as an explosion, through the earth until they encounter a reflecting surface and are reflected into a detector, such as a geophone. Figure 4-1 shows a seismic trace. Each trace represents the signal received by a detector. Changes to the amplitude

24

of seismic waves occur at reflectors. A seismic reflection occurs at the interface between two regions with different acoustic impedances.

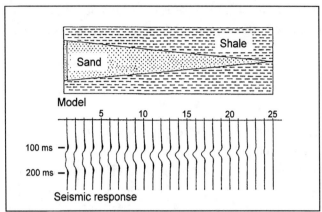

Figure 4-1. Seismic trace for a sand wedge (after Ruijtenberg, 1990; reprinted by permission of the Society of Petroleum Engineers).

Acoustic impedance is a fundamental seismic parameter. Acoustic impedance is defined as $Z = \rho V$ where ρ is the bulk density of the medium and V is the compressional velocity of the wave in the medium. Figure 4-2 illustrates a correlation between seismic wave velocity and the bulk density of different types of rock. Further discussion of rock properties and their relationship to seismic variables can be found in the literature [for example, Schön 1996].

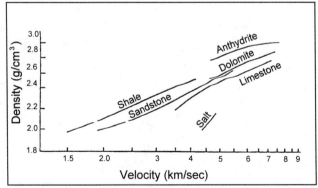

Figure 4-2. Seismic wave velocity and bulk density of rock (after Telford, et al., 1976; reprinted by permission of Cambridge University Press; after Gardner, et al., 1974).

A change in acoustic impedance will cause a reflection of the sound wave. The ability to reflect a sound wave by a change in acoustic impedance is quantified in terms of the reflection coefficient. The reflection coefficient R at the interface between two contiguous layers is defined in terms of acoustic impedances as

$$R = \frac{Z_2 - Z_1}{Z_2 + Z_1} = \frac{\rho_2 V_2 - \rho_1 V_1}{\rho_2 V_2 + \rho_1 V_1}$$

where subscripts 1 and 2 refer to the contiguous layers.

Reflection coefficient magnitudes for typical subsurface interfaces are illustrated in Table 4-1. Values of reflection coefficients at the sandstone/limestone interface show that reflection coefficient values can be relatively small. In addition to reflection coefficient, a transmission coefficient can be defined. The transmission coefficient is one minus the reflection coefficient.

Table 4-1

Typical Reflection Coefficients

Interface	Reflection Coefficient
Sandstone on limestone	0.040
Limestone on sandstone	- 0.040
Ocean bottom	0.11 (soft) to 0.44 (hard)

Nonzero reflection coefficients occur when a wave encounters a change in acoustic impedance, either because of a change in compressional velocity of the wave as it propagates from one medium to another, or because the bulk densities of the media differ. If the change in acoustic impedance is large enough, the reflection can be measured at the surface. That is why gas tends to show up as bright spots on seismic data – there is a big change in the density of the fluid. By contrast, the presence of an oil/water contact is harder to observe with seismic measurements because density differences between the oil and water phases are relatively small and result in small changes in acoustic impedance.

The seismic trace plots seismic amplitude versus two-way travel time, or the time it takes the seismic wave to propagate from the source to the receiver.

One of the central problems in seismic data processing is to determine the time/depth conversion. The conversion of travel time data to formation depth requires that the velocity associated with each geologic zone be known or can be inferred as the wave evolves with time. When the time/depth conversion is applied to seismic data, it can change the relative depths of seismic amplitudes associated with adjacent traces.

Figure 4-3 shows the amplitude and wavelength of a seismic wave [after de Buyl, et al., 1988]. The sonic log response shown in Figure 4-3 illustrates the relationship between seismic amplitude and the sonic log. Sonic logs are typically used to calibrate seismic data when seismic data is used in reservoir characterization. The sonic log response in Figure 4-3 delineates the top and base of a geologic section.

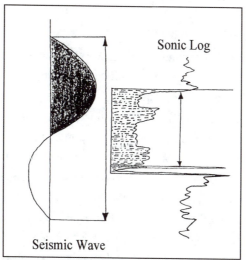

Figure 4-3. Seismic wave and sonic log response.

The wavelength of the seismic wave is the velocity of the wave divided by its frequency. Alternatively, the wavelength is the velocity in a given medium times the period of the wave. The frequency of the wave is a measure of the energy of the wave and is conserved as the wave propagates from one medium to another. The wavelength, however, can vary from one medium to another.

When waves overlap – or superpose – they create a wavelet, as shown in Figure 4-4. The time duration associated with the wavelet disturbance is denoted Δt. The wavelet has a velocity V in a medium, and the period T is the width of the wavelet when plotted as a trace on a time-map of seismic data. The length of the wave is equal to the velocity V times the period T. Thus, if the wavelet has a 10 millisecond period and the velocity is 5000 feet per second in a particular medium, then the length L of that wavelet is 50 feet.

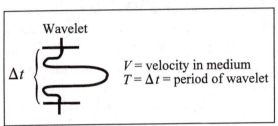

Figure 4-4. Seismic wavelet.

If seismic data has enough resolving power to show the reflecting boundaries of a geologic layer, then the amplitudes of the seismic waves may be useful for further characterizing petrophysical properties of the reservoir. For example, suppose a reservoir region is characterized by a porosity ϕ, permeability K, net thickness h_{net}, and oil saturation S_o. Seismic amplitude may be correlatable with rock quality (for example, Kh_{net} or ϕkh_{net}) or oil productive capacity (for example, $S_o \phi kh_{net}$). When a correlation does exist between seismic amplitude and a grouping of petrophysical parameters, the correlation may be used to help guide the distribution of reservoir properties in areas between wells.

Figures 4-5a and b show two approaches to contouring a set of values at control points. The smooth contour lines shown in Figure 4-5a are preferred by mappers [Tearpock and Bischke, 1991] unless the undulating contour lines in Figure 4-5b are supported by additional data. Seismic correlations can be used to justify the more heterogeneous contouring style shown in Figure 4-5b. A growing body of literature provides additional discussion of this application in the context of an emerging discipline known as reservoir geophysics. For example, see de Buyl, et al.[1988], Evans [1996], Blackwelder, et al. [1996], and Beasley [1996]. Reservoir geophysics is discussed further in Chapters 12 and 30.

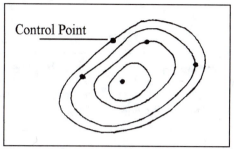

Figure 4-5a. Smooth contour lines.

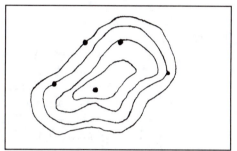

Figure 4-5b. Undulating contour lines.

4.2 Mega Scale

The Giga Scale helps define reservoir architecture, but is too coarse to provide the detail needed to design a reservoir development plan. The Mega Scale is the scale at which we begin to integrate well log and well test data into a working model of the reservoir. Table 4-2 illustrates the type of information that can be obtained at the Mega Scale level from well log data. The most common interpretations of each log response are included in the table. For example, a high gamma ray response implies the presence of shales, while a low gamma ray response implies the presence of clean sands or carbonates. A combination of well logging tools is usually needed to minimize ambiguity in log interpretation, as discussed by Brock [1986].

Table 4-2

Well Log Response

Log	Variable	Response
Gamma ray	Rock type	Detects shale from in situ radioactivity. ♦ High GR ⇒ shales ♦ Low GR ⇒ clean sands or carbonates
Resistivity	Fluid type	Measures resistivity of formation water. ♦ High resistivity ⇒ hydrocarbons ♦ Low resistivity ⇒ brine
Density	Porosity	Measures electron density by detecting Compton scattered gamma rays. Electron density is related to formation density. Good for detecting hydrocarbon gas with low density compared to rock or liquid. ♦ Low response ⇒ low HC gas content ♦ Large response ⇒ high HC gas content
Acoustic (sonic)	Porosity	Measures speed of sound in medium. Speed of sound is faster in rock than in fluid. ♦ Long travel time ⇒ slow speed ⇒ large pore space ♦ Short travel time ⇒ high speed ⇒ small pore space
Neutron	Hydrogen content	Fast neutrons are slowed by collisions to thermal energies. Thermal neutrons are captured by nuclei, which then emit detectable gamma rays. Note: Hydrogen has a large capture cross-section for thermal neutrons. Good for detecting gas. ♦ Large response ⇒ high H content ♦ Small response ⇒ low H content
Spontaneous potential	Permeable beds	Measures electrical potential (voltage) associated with movement of ions. ♦ Low response ⇒ impermeable shales ♦ Large response ⇒ permeable beds

Table 4-3 from Kamal, et al. [1995] illustrates the type of information that can be obtained at the Mega Scale level from well test data. The table also notes the time in the life of the project when the well test is most likely to be run. It is usually necessary to run a variety of well tests as the project matures. These tests help refine the operator's understanding of the field and often motivate changes in the way the well or the field is operated. Additional information about well testing can be found in literature sources such as Matthews and Russell [1967], Earlougher [1977], and Sabet [1991].

Table 4-3
Reservoir Properties Obtainable from Transient Tests

Type of Test	Properties	Development Stage
Drill stem tests	Reservoir behavior Permeability Skin Fracture length Reservoir pressure Reservoir limit Boundaries	Exploration and appraisal wells
Repeat-formation Tests / Multiple formation tests	Pressure profile	Exploration and appraisal wells
Drawdown tests	Reservoir behavior Permeability Skin Fracture length Reservoir limit Boundaries	Primary, secondary and enhanced recovery
Buildup tests	Reservoir behavior Permeability Skin Fracture length Reservoir pressure Reservoir limit Boundaries	Primary, secondary and enhanced recovery

Table 4-3 (continued)
Reservoir Properties Obtained from Transient Tests

Type of Test	Properties	Development Stage
Step-rate tests	Formation parting pressure Permeability Skin	Secondary and enhanced recovery
Falloff tests	Mobility in various banks Skin Reservoir pressure Fracture length Location of front Boundaries	Secondary and enhanced recovery
Interference and pulse tests	Communication between wells Reservoir type behavior Porosity Interwell permeability Vertical permeability	Primary, secondary and enhanced recovery
Layered reservoir tests	Properties of individual layers Horizontal permeability Vertical permeability Skin Average layer pressure Outer boundaries	Throughout reservoir life

Tables 4-2 and 4-3 illustrate a few of the methods used to gather Mega Scale information. Advances in technology periodically add to a growing list of transient tests and well log tools [for example, see Kamal, 1995; Felder, 1994]. In many cases, budgetary constraints will be the controlling factor in determining the number and type of tests run. The modeling team must work with whatever information is available. Occasionally, an additional well test or well log will need to be run, but the expense and scheduling make it difficult to justify acquiring new well log or well test information once a simulation study is underway.

Exercises

Exercise 4.1 Seismic Parameters: Data set EXAM11.DAT is a cross-section model of a two-layer gas reservoir undergoing depletion with aquifer influx into the lower layer. Run EXAM11.DAT and find the initial compressional velocity (V_P), reflection coefficient (RC), and ratio of compressional velocity to shear velocity (V_p/V_s). Activate maps IVPMAP, IRCMAP and IVRMAP using the information given in Chapter 20.1. Notice how V_P, RC, and V_p/V_s change in block I = 1 of layer K = 2 (the lower layer) as water moves into the layer.

Exercise 4.2 Repeat Exercise 4.1 using a critical gas saturation of 0. This should be achieved by setting the relative permeability of gas to 0.01 at a gas saturation of 0.02.

Exercise 4.3 Repeat Exercise 4.1 using a grain bulk modulus that is equal to the frame bulk modulus.

References

Anderson, R.N. (1995): "Method Described for Using 4D Seismic to Track Reservoir Fluid Movement," *Oil & Gas Journal*, pp. 70-74, April 3.

Ausburn, B.E., A.K. Nath, and T.R. Wittick (Nov. 1978): "Modern Seismic Methods – An Aid for the Petroleum Engineer," *Journal of Petroleum Technology*, pp. 1519-1530.

Beasley, C.J. (1996): "Seismic Advances Aid Reservoir Description," *Journal of Petroleum Technology*, pp. 29-30.

Blackwelder, B., L. Canales, and J. Dubose (1996): "New Technologies in Reservoir Characterization," *Journal of Petroleum Technology*, pp. 26-27.

Brock, J. (1986): **Applied Open-Hole Log Analysis**, Houston: Gulf Publishing.

de Buyl, M., T. Guidish, and F. Bell (1988): "Reservoir Description from Seismic Lithologic Parameter Estimation," *Journal of Petroleum Technology*, pp. 475-482.

Earlougher, R.C., Jr. (1977): **Advances in Well Test Analysis**, SPE Monograph Series, Richardson, TX: Society of Petroleum Engineers.

Evans, W.S. (1996): "Technologies for Multidisciplinary Reservoir Characterization," *Journal of Petroleum Technology*, pp. 24-25.

Felder, R.D. (1994): "Advances in Openhole Well Logging," *Journal of Petroleum Technology*, pp. 693-701.

Gardner, H.F., L.W. Gardner, and A.H. Gregory (1974): "Formation Velocity and Density – the Diagnostic Basis for Stratigraphic Traps," *Geophysics* Volume 39, pp. 770-780.

He, W., R.N. Anderson, L. Xu, A. Boulanger, B. Meadow, and R. Neal (1996): "4D Seismic Monitoring Grows as Production Tool," *Oil & Gas Journal*, pp. 41-46, May 20.

Johnston, D.H. (1997): "A Tutorial on Time-Lapse Seismic Reservoir Monitoring," *Journal of Petroleum Technology*, pp. 473-475.

Kamal, M.M., D.G. Freyder, and M.A. Murray (1995): "Use of Transient Testing in Reservoir Management," *Journal of Petroleum Technology*, pp. 992-999.

Matthews, C.S. and D.G. Russell (1967): **The Reservoir Engineering Aspects of Waterflooding**, SPE Monograph Series, Richardson, TX: Society of Petroleum Engineers.

McQuillin, R., M. Bacon, and W. Barclay (1984): **An Introduction to Seismic Interpretation**, Houston: Gulf Publishing.

Richardson, J.G. (Feb. 1989): "Appraisal and Development of Reservoirs," *Geophysics: The Leading Edge of Exploration*, pp. 42 ff.

Ruijtenberg, P.A., R. Buchanan, and P. Marke (1990): "Three-Dimensional Data Improve Reservoir Mapping," *Journal of Petroleum Technology*, pp. 22-25, 59-61.

Sabet, M.A. (1991): **Well Test Analysis**, Houston: Gulf Publishing.

Schön, J.H. (1996): **Physical Properties of Rocks: Fundamentals and Principles of Petrophysics**, New York: Elsevier, Vol. 18.

Sheriff, R.E. (1989): **Geophysical Methods**, Englewood Cliffs, NJ: Prentice-Hall.

Tearpock, D.J. and R.E. Bischke (1991): **Applied Subsurface Geological Mapping**, Englewood Cliffs, NJ: Prentice Hall.

Telford, W.M., L.P. Geldart, R.E. Sheriff, and D.A. Keys (1976): **Applied Geophysics**, Cambridge: Cambridge University Press.

Chapter 5

Fluid Properties

Properties of petroleum fluids must be quantified in a reservoir simulator. The range of applicability of a reservoir simulator is defined, in part, by the types of fluids that can be modeled using the mathematical algorithms coded in the simulator. For these reasons, it is worth considering the general types of fluids that may be encountered in a commercial reservoir environment [for example, see Pedersen, et al., 1989; Koederitz, et al., 1989, McCain, 1973; and Amyx, et al., 1960].

5.1 Fluid Types

An estimate of the elemental composition (by mass) of petroleum is given in the following chart:

Carbon	84% - 87%
Hydrogen	11 % - 14%
Sulphur	0.6% - 8 %
Nitrogen	0.02% - 1.7%
Oxygen	0.08% - 1.8%
Metals	0% - 0.14%

It can be seen from the table that petroleum fluids are predominantly hydrocarbons. The most common hydrocarbon molecules are paraffins, napthenes, and aromatics because of the relative stability of the molecules. A paraffin is a saturated hydrocarbon, that is, it has a single bond between carbon atoms. Examples include methane and ethane. Paraffins have the general chemical formula C_nH_{2n+2}. Napthenes are saturated hydrocarbons with a ringed structure, as in cyclopentane. They have the general chemical formula $C_n H_{2n}$. Aromatics are unsaturated hydrocarbons with a ringed structure that have multiple bonds between the carbon atoms as in benzene. The unique ring structure makes aromatics relatively stable and unreactive.

A general PVT diagram of a pure substance displays phase behavior as a function of pressure, volume and temperature. The types of properties of interest from a reservoir engineering perspective can be conveyed in a pressure-temperature (P-T) diagram of phase behavior like the one shown in Figure 5-1 (after Craft and Hawkins [1959]). Most reservoir fluids do not exhibit significant temperature effects in situ, although condensate reservoirs in thick sands may display a compositional gradient that can influence yield as a function of well perforation depth.

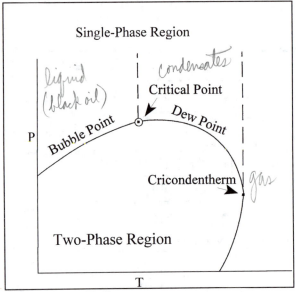

Figure 5-1. P-T diagram.

The P-T diagram includes both single-phase and two-phase regions. The line separating the single-phase region from the two-phase region is called the phase envelope. The black oil region is at low temperature and in the high pressure region above the bubble point curve separating the single-phase and two-phase regions. If we consider pressures in the single-phase region and move to the right of the diagram by letting temperature increase towards the critical point, we encounter volatile oils. At temperatures above the critical point but less than the cricondentherm, reservoir fluids behave like condensates. The cricondentherm is the maximum temperature at which a fluid can exist in both the gas and liquid phases. When reservoir temperature is greater than the cricondentherm, we encounter gas reservoirs. A summary of these fluid types is given in Table 5-1. Notice that separator gas-oil ratio (GOR) is a useful indicator of fluid type.

Table 5-1
Rules of Thumb for Classifying Fluid Types

Fluid Type	Separator GOR (MSCF/STB)	Pressure Depletion Behavior in Reservoir
Dry gas	No surface liquids	Remains gas
Wet gas	> 100	Remains gas
Condensate	3 - 100	Gas with liquid drop out
Volatile oil	1.5 - 3	Liquid with significant gas
Black oil	0.1 - 1.5	Liquid with some gas
Heavy oil	~ 0	Negligible gas formation

Let us consider a reservoir containing hydrocarbons that are at a pressure and temperature corresponding to the single-phase black oil region. If reservoir pressure declines at constant temperature, the reservoir pressure will eventually cross the bubble point pressure curve and enter the two-phase gas-oil region. Similarly, starting with a single-phase condensate and letting reservoir pressure decline at constant temperature, the reservoir pressure will cross the dew point pressure curve to enter the two-phase region. In this case, a free phase liquid drops out of the condensate gas. Once liquid drops out, it is very difficult to recover. One recovery method is dry gas cycling, but the recovery efficiency will be substantially less than 100%. If we drop the pressure even further, it is possible to encounter retrograde condensation for some hydrocarbon compositions.

The P-T diagram also applies to temperature and pressure changes in a wellbore. In the case of wellbore flow, the fluid moves from relatively high reservoir temperature and pressure to relatively low surface temperature and pressure. As a result, it is common to see fluids that are single-phase in the reservoir become two-phase by the time they reach the surface.

Figure 5-2 is a P-T diagram that compares two-phase envelopes for four types of fluids. A reservoir fluid can change from one fluid type to another depending on how the reservoir is produced. A good example is dry gas injection into a black oil reservoir. Dry gas injection increases the relative amount of low molecular weight components in the black oil. The two-phase envelope rotates counter-clockwise in the P-T diagram as the relative amount of lower molecular weight components increases. Similarly, dry gas injection into a condensate can make the phase envelope transform from one fluid type to another. Thus, the way the reservoir is operated has a significant impact on fluid behavior.

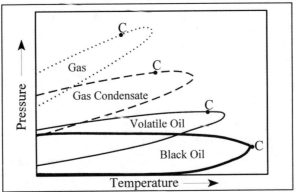

Figure 5-2. Typical two-phase P-T envelopes for different fluid types.

Table 5-2 shows different compositions for typical fluid types. Dry gas usually contains only the lower molecular weight components. Gas condensates start to add higher molecular weight components. Volatile oils continue to add higher molecular weight components. The addition of higher molecular weight components and the reduction of lower molecular weight components eventually yields a black oil. If we monitor methane content (C_1), we see that it tends to decrease as fluids change from dry gas to black oil.

light → **Table 5-2**
Typical Molar Compositions of Petroleum Fluid Types
heavy →
[after Pedersen, et al., 1989]

Component	Gas	Gas Condensate	Volatile Oil	Black Oil
N_2	0.3	0.71	1.67	0.67
CO_2	1.1	8.65	2.18	2.11
C_1	90.0	70.86	60.51	34.93
C_2	4.9	8.53	7.52	7.00
C_3	1.9	4.95	4.74	7.82
$iC_4 + nC_4$	1.1	2.00	4.12	5.48
$iC_5 + nC_5$	0.4	0.81	2.97	3.80
$iC_6 + nC_6$	C_{6+}: 0.3	0.46	1.99	3.04
C_7		0.61	2.45	4.39
C_8		0.71	2.41	4.71
C_9		0.39	1.69	3.21
C_{10}		0.28	1.42	1.79
C_{11}		0.20	1.02	1.72
C_{12}		0.15	C_{12+}: 5.31	1.74
C_{13}		0.11		1.74
C_{14}		0.10		1.35
C_{15}		0.07		1.34
C_{16}		0.05		1.06
C_{17}		C_{17+}: 0.37		1.02
C_{18}				1.00
C_{19}				0.90
C_{20}				C_{20+}: 9.18

5.2 Fluid Modeling

In general, fluid behavior is best modeled using an equation of state. Table 5-3 shows some cubic equations of state (EoS) used in commercial compositional simulators. In addition to pressure (P), volume (V), and temperature

(T), the EoS contains the gas constant R and a set of adjustable parameters {a, b} which may be functions of temperature. The EoS in Table 5-3 are called "cubic" because they yield a cubic equation for the compressibility factor $Z = PV/RT$. In the case of an ideal gas, $Z = 1$.

Table 5-3

Examples of Cubic Equations of State

Redlich-Kwong	$P = \dfrac{RT}{V-b} - \dfrac{a/T^{1/2}}{V(V+b)}$
Soave-Redlich-Kwong	$P = \dfrac{RT}{V-b} - \dfrac{a(T)}{V(V+b)}$
Peng-Robinson	$P = \dfrac{RT}{V-b} - \dfrac{a(T)}{V(V+b) + b(V-b)}$
Zudkevitch-Joffe	$P = \dfrac{RT}{V-b(T)} - \dfrac{a(T)/T^{1/2}}{V[V+b(T)]}$

Equations of state are valuable for representing fluid properties in many situations. For example, suppose we want to model a system in which production is commingled from more than one reservoir with more than one fluid type. In this case the most appropriate simulator would be a compositional simulator because a black oil simulator would not provide as accurate a representation of fluid behavior.

The two most common types of reservoir fluid models are black oil models and compositional models. Black oil models are based on the assumption that the saturated phase properties of two hydrocarbon phases (oil and gas) depend on pressure only. Compositional models also assume two hydrocarbon phases, but they allow the definition of many hydrocarbon components. Unlike a black oil simulator, which can be thought of as a compositional simulator with two components, a compositional simulator often has six to ten components. By comparison, process engineering simulators that are used to model surface facilities typically require up to 20 components or more. The cost of running a compositional simulator increases dramatically with increases in the number of components modeled, but the additional components make it possible to more accurately model complex fluid phase behavior. If compositional model results are to be used in a process engineering model, it is often necessary to compromise on the number of components to be used for each application.

Equations of state must be used to calculate equilibrium relations in a compositional model. This entails tuning parameters such as EoS parameters $\{a, b\}$ in Table 5-3. Several regression techniques exist for tuning an EoS. They usually differ in the choice of EoS parameters that are to be varied in an attempt to match lab data with the EoS.

Figures 5-3 and 5-4 show typical fluid property behavior of gas and oil properties for a black oil model. Gas phase properties are gas formation volume factor (B_g), gas viscosity (μ_g), and liquid yield (r_s). Oil phase properties are oil formation volume factor (B_o), oil viscosity (μ_o), and solution GOR (R_{so}). Both saturated and undersaturated curves are included as functions of pressure only. Phase changes occur at the saturation pressures. Single-phase oil becomes two-phase gas-oil when pressure drops below the bubble point pressure (P_b), and single-phase gas becomes two-phase gas condensate when pressure drops below the dew point pressure (P_d).

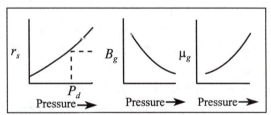

Figure 5-3. Gas phase properties.

Simulators run most efficiently when fluid property data are smooth curves. Any discontinuity in a curve can cause numerical difficulties. Ordinarily, realistic fluid properties are smooth functions of pressure except at points where phase transitions occur. As a practical matter, it is usually wise to plot input PVT data to verify the smoothness of the data. Most simulators reduce the nonlinearity of the gas formation volume factor B_g by using the inverse $b_g = 1/B_g$ to interpolate gas properties.

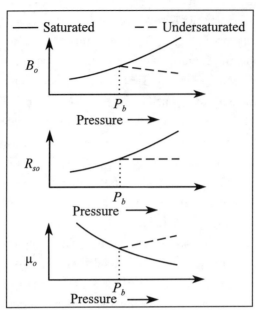

Figure 5-4. Oil phase properties.

Oil properties from a laboratory must usually be corrected for use in a black oil simulator [Moses, 1986]. Flow in the reservoir is a slow process corresponding to a differential process in the laboratory. When oil is produced, however, it is flashed to the surface through several pressure regimes. The corrections are designed to more adequately represent fluids as they flow differentially in the reservoir prior to being flashed to surface conditions. The corrections alter solution gas-oil ratio and oil formation volume factor. The effect of the correction is illustrated for the case study in Chapter 13.5. The oil property correction is often significant.

Water properties must also be entered in a simulator. Ideally water properties should be measured by performing laboratory analyses on produced water samples. If samples are not available, correlations are often sufficiently accurate for describing the behavior of water.

In the absence of reliable fluid data for any of the reservoir fluids, it may be necessary to use correlations. McCain [1991] reviewed the state of the art in the use of correlations to describe fluid properties.

5.3 Fluid Sampling

All laboratory measurements of fluid properties and subsequent analyses are useless if the fluid samples do not adequately represent *in situ* fluids. The goal of fluid sampling is to obtain a sample that is representative of the original fluid in the reservoir. It is often necessary to condition the well before the sample is taken. A well is conditioned by producing any nonrepresentative fluid, such as drilling mud, from within and around the wellbore until it is replaced by original reservoir fluid flowing into the wellbore. Fluid samples may then be taken from either the surface or subsurface.

Subsurface sampling requires lowering a pressurized container to the production interval and subsequently trapping a fluid sample. This is routinely accomplished by drill stem testing, especially when access to surface facilities is limited. It is generally cheaper and easier to take surface samples from separator gas and oil.

If a surface sample is taken, the original *in situ* fluid, that is, the fluid at reservoir conditions, must then be reconstituted by combining separator gas and separator oil samples. The recombination step assumes accurate measurements of flow data at the surface, especially gas oil ratio. Subsurface sampling from a properly conditioned well avoids the recombination step, but is more difficult and costly than surface sampling, and usually provides a smaller volume of sample fluid. The validity of fluid property data depends on the quality of the fluid sampling procedure.

Exercises

Exercise 5.1 Data set EXAM9.DAT models depletion of a gas reservoir with aquifer support. Initial reservoir pressure is approximately 1947 psia. By looking at reservoir pressure and gas viscosity at 2015 psia, describe the effect of reducing reservoir temperature from 226°F to 150°F on the behavior of the model. For this example, neglect the temperature dependence of water properties. Refer to Chapter 19.6 for a description of BOAST4D fluid property input data.

Exercise 5.2 Data file CS-VC4.DAT is a vertical column model with four layers. Layers K = 1, 3, 4 are pay zones, and layer K = 2 is a shale layer. The data set is a model of primary depletion of an initially undersaturated oil reservoir. Run CS-VC4.DAT for three years and see how gas saturation changes as pressure declines. You should see gravity segregation and the formation of a gas cap in layer K = 3. By referring to Chapter 20.2 and file BTEMP.WEL, determine which model layers are being depleted through wellbore perforations.

Exercise 5.3 Determine the effect on gas cap and reservoir pressure when solution gas-oil ratio in CS-VC4.DAT is replaced with the following data. Run the modified data set for a period of three years, and then compare the results with the results of Exercise 5.2.

Pressure (psia)	Solution Gas-Oil Ratio (SCF/STB)
14.7	1.0
514.7	54.0
1014.7	105.0
1514.7	209.0
2014.7	292.0
2514.7	357.0
3014.7	421.0
4014.7	486.0
5014.7	522.0
6014.7	550.0

Exercise 5.4 Run the data set prepared in Exercise 5.3 with the assumption that no fluids can flow between model layers (multiply z direction transmissibility by zero).

Exercise 5.5 Run CS-VC4.DAT with the bubble point pressure reduced by 500 psia. What effect does this have on solution gas-oil ratio and model performance?

References

Amyx, J.W., D.H. Bass, and R.L. Whiting (1960): **Petroleum Reservoir Engineering**, New York: McGraw-Hill.

Craft, B.C. and M.F. Hawkins (1959): **Applied Petroleum Reservoir Engineering**, Englewood Cliffs, NJ: Prentice-Hall.

Koederitz, L.F., A.H. Harvey, and M. Honarpour (1989): **Introduction to Petroleum Reservoir Analysis**, Houston: Gulf Publishing.

McCain, W.D., Jr. (1973): **The Properties of Petroleum Fluids**, Tulsa, OK: Petroleum Publishing.

McCain, W.D., Jr. (1991): "Reservoir-Fluid Property Correlations – State of the Art," *Society of Petroleum Engineers Reservoir Engineering*, pp. 266-272.

Moses, P.L. (July 1986): "Engineering Applications of Phase Behavior of Crude Oil and Condensate Systems," *Journal of Petroleum Technology*, pp. 715-723; and F.H. Poettmann and R.S. Thompson (1986): "Discussion of Engineering Applications of Phase Behavior of Crude Oil and Condensate Systems," *Journal of Petroleum Technology*, pp. 1263-1264.

Pedersen, K.S., A. Fredenslund, and P. Thomassen (1989): **Properties of Oil and Natural Gases**, Houston: Gulf Publishing.

Chapter 6

Rock-Fluid Interaction

The previous two chapters described the data needed to model the solid structure of the reservoir and the behavior of fluids contained within the solid structure. Small scale laboratory measurements of fluid flow in porous media show that fluid behavior depends on the properties of the solid material. The interaction between rock and fluid is modeled using a variety of physical parameters that include relative permeability and capillary pressure [Collins, 1961; Dake, 1978; Koederitz, et al., 1989]. Laboratory measurements provide information at the core scale (Macro Scale) and, in some cases, at the microscopic scale (Micro Scale). They are the subject of the present chapter.

6.1 Porosity, Permeability, Saturation and Darcy's Law

Porosity, permeability and saturation can be obtained from Mega Scale measurements such as well logs and well tests, and by direct measurement in the laboratory. Comparing values of properties obtained using methods at two different scales demonstrates the sensitivity of important physical parameters to the scale at which they were measured. Ideally there will be good agreement between the two scales, that is, well log porosity or well test permeability will agree with corresponding values measured in the laboratory. In many cases, however, there are disagreements. Assuming measurement error is not the source of disagreement, differences in values show that differences in scale can impact the measured value of the physical parameter. A well test permeability, for example, represents an average over an area of investigation that is very large compared to a laboratory measurement of permeability using a six inch core sample. The modeling team often has to make judgements about the relative merits of contradictory data. The history matching process recognizes this source of uncertainty, as is discussed in subsequent chapters.

The most common types of reservoir rock are listed in Table 6-1. One of the most fundamental properties of rock that must be included in a reservoir model is porosity. Porosity is the fraction of a porous medium that is void space. If the void space is connected and communicates with a wellbore, it is referred to as effective porosity, otherwise the void space is ineffective porosity. The original porosity resulting from sediment deposition is called primary porosity. Secondary porosity is an incremental increase in primary porosity due to the chemical dissolution of reservoir rocks, especially carbonates. Primary and secondary porosity can be both effective and ineffective. Total porosity is a combination of ineffective porosity and effective (interconnected) porosity.

Table 6-1
Common Reservoir Rocks

Sandstones	Compacted sediment Conglomerate
Shales	Laminated sediment Predominantly clay
Carbonates	Produced by chemical and biochemical sources Limestone

Porosity values depend on rock type, as shown in Table 6-2. There are two basic techniques for directly measuring porosity: core analysis in the laboratory, and well logging. Laboratory measurements tend to be more accurate, but sample only a small fraction of the reservoir. Changes in rock properties may also occur when the core is brought from the reservoir to the surface. Well log measurements sample a much larger portion of the reservoir than core analysis, but typically yield less accurate values. Ideally, a correlation can be established between *in situ* measurements such as well logging, and surface measurements such as core analysis.

Table 6-2
Dependence of Porosity on Rock Type

Rock Type	Porosity Range (%)	Typical Porosity (%)
Sandstone	15-35	25
Unconsolidated sandstone	20-35	30
Carbonate • Intercrystalline limestone • Oolitic limestone • Dolomite	 5-20 20-35 10-25	 15 25 20

Darcy's Law is the basic equation describing fluid flow in a simulator. Darcy's equation for single phase flow is

$$Q = -0.001127 \frac{KA}{\mu} \frac{\Delta P}{\Delta x}$$

where the physical variables are defined in oil field units as

Q = flow rate (bbl/day)

A = cross-sectional area (ft^2)

μ = fluid viscosity (cp)

K = permeability (md)

P – pressure (psi)

x = length (ft)

Darcy's Law says that rate is proportional to cross-sectional area times pressure difference ΔP across a distance Δx, and is inversely proportional to the viscosity of the fluid. The minus sign shows that the direction of flow is opposite to the direction of increasing pressure; fluids flow from high pressure to low pressure in a horizontal (gravity-free) system.

The linearity of Darcy's Law is an approximation that is made by virtually all commercial simulators. Fluid flow in a porous medium can have a nonlinear effect that is represented by the Forcheimer equation [Govier, 1978]. The nonlinear effect becomes more important in high flow rate gas wells.

Permeability is a physical constant describing flow in a given sample for a given fluid and set of experimental conditions. If those conditions are changed, the permeability being measured may not apply. For example, if a waterflood is planned for a reservoir that is undergoing gravity drainage, laboratory measured permeabilities need to represent the injection of water into a core with hydrocarbon and connate water. The permeability distribution and the relative permeability curves put in the model need to reflect the type of processes that occur in the reservoir.

Permeability has meaning as a statistical representation of a large number of pores. A Micro Scale measurement of grain size distribution shows that different grain sizes and shapes affect permeability. Permeability usually decreases as grain size decreases. It may be viewed as a mathematical convenience for describing the statistical behavior of a given flow experiment. In this context, transient testing gives the best measure of permeability over a large volume. Despite its importance to the calculation of flow, permeability and its distribution will not be known accurately. Seismic data can help define the distribution of permeability between wells if a good correlation exists between seismic amplitude and a rock quality measurement that includes permeability.

It is not unusual to find that permeability has a directional component: that is, permeability is larger in one direction than another [for example, see Fanchi, et al., 1996]. When a model is being designed, the modeling team should account for the direction associated with permeability. In principle, simulators can take all of these effects into account. In practice, however, the tensor permeability discussed in the literature by, for example, Bear [1972], Lake [1988], and Ekrem and Lake [1990] is seldom reflected in a simulator. The usual assumption is that permeability is aligned along one of three orthogonal directions known as the principal axes of the tensor. This assumption has implications for model studies that should be considered when assessing model results (see Chapter 8 and Fanchi [1983]).

In many cases vertical permeability is not measured and must be assumed. A rule of thumb is to assume vertical permeability is approximately one tenth of horizontal permeability. These are reasonable assumptions when there is no data to the contrary.

6.2 Relative Permeability and Capillary Pressure

Reservoir models calculate saturation as a function of time. Consider the case of water displacing oil. Initially, oil occupies the interior of pore spaces and connate water is adjacent to the rock surface of a water-wet reservoir. When the flood begins, water displaces oil through the interconnected pore space. The measure of interconnectedness is permeability. The oil left behind after the waterflood is residual or irreducible oil saturation. Similar behavior is seen for other combinations of multiphase flow, for example, gas-oil, gas-water and gas-oil-water. Multiphase flow is modeled by including relative permeability curves in the simulator. Saturation end points for the relative permeability curves are used to establish initial fluids-in-place in addition to modeling flow behavior.

A typical set of relative permeability curves is shown in Figure 6-1. Relative permeability curves represent flow mechanisms, such as drainage or imbibition processes, or fluid wettability. Relative permeability data should be obtained by experiments that best model the type of displacement that is thought to dominate reservoir flow performance. For example, water-oil imbibition curves are representative of waterflooding, while water-oil drainage curves describe the movement of oil into a water zone. The modeling team needs to realize that the relative permeability curves used in a flow model are most representative of the type of experiment that was used to measure the curves. Applying these curves to another type of displacement mechanism can introduce significant error.

Several procedures exist for averaging relative permeability data [for example, Schneider, 1987; Mattax and Dalton, 1990]. In practice, relative permeability is one of the most useful physical quantities available for performing a history match. As a consequence, the curves that are initially entered into a reservoir model are often modified during the history match process. In the absence of measured data, correlations such as Honarpour, et al. [1982] give a reasonable starting point for estimating relative permeability.

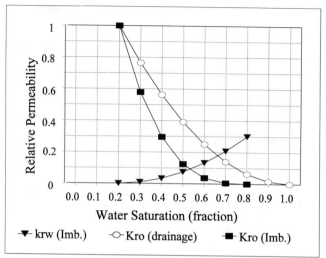

Figure 6-1. Typical water-oil relative permeability curves.

Capillary pressure is usually included in reservoir simulators. The relationship between capillary pressure and elevation is used to establish the initial transition zone in the reservoir. The oil-water transition zone, for example, is the zone between water-only flow and oil-only flow. It represents that part of the reservoir where 100% water saturation grades into oil saturation with irreducible water saturation. Similar transition zones may exist at the interface between any pair of immiscible phases.

Capillary pressure data is used primarily for determining initial fluid contacts and transition zones. It is also used in fractured reservoir models for controlling the flow of fluids between the fracture and the rock matrix. If capillary pressure is neglected, transition zones are not included in the model. This is illustrated in Figure 6-2. Figure 6-3 shows the effect of neglecting capillary pressure when a grid is used to represent the reservoir. The fluid content of the block is determined by the location of the block mid-point relative to a contact between two phases. The block mid-point is shown as a dot in the center of the blocks in Figure 6-3. Thus, if the block mid-point is above the gas-oil contact (GOC), the entire block is treated as a gas cap block (single phase gas with irreducible water saturation), even if much of the block extends into the oil column. A more accurate representation may be obtained by decreasing the thickness of the grid blocks, but this often results in a substantial increase in the cost of making computer runs. The relative benefits of incremental accuracy versus incremental cost must be considered when modeling transition zones.

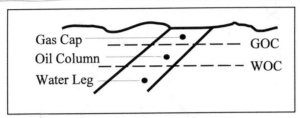

Figure 6-2. Case 1: Neglect transition zones.

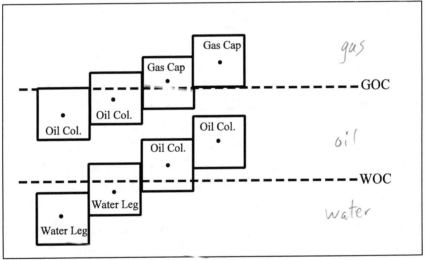

Figure 6-3. Initial fluid distribution in model without transition zone.

The inclusion of a transition zone in the model requires specifying a capillary pressure (P_c) curve as a function of saturation for whatever transition zone is being modeled: oil-water, gas-oil, or gas-water. The height h_{tz} of the transition zone above the free water level (the level corresponding to $P_c = 0$ psia) is proportional to the capillary pressure and inversely proportional to the density difference ($\Delta\rho$) between the two fluids:

$$h_{tz} = \frac{P_c}{\Delta\rho}$$

The height of the transition zone is a function of saturation because capillary pressure depends on saturation. The oil-water transition zone is typically the thickest transition zone because the density difference between oil and water is less than the density difference between gas and an immiscible liquid.

Figures 6-4 and 6-5 illustrate the initialization of a model containing a nonzero capillary pressure curve. First, the height h_{tz} above a specified contact, such as the water-oil contact (WOC), is calculated from P_c and $\Delta\rho$. The saturation of a block with a mid-point at height h_{tz} above the contact is then calculated from the relationship between capillary pressure and saturation.

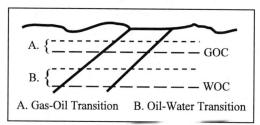

Figure 6-4. Case 2: Include transition zone in model.

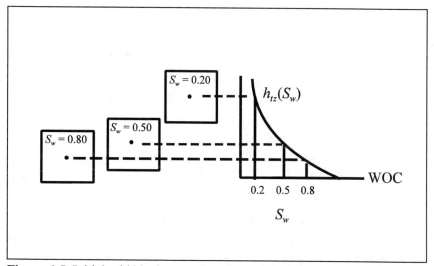

Figure 6-5. Initial grid block saturations in model with transition zone.

Transition zones complicate the identification of fluid contacts because the definition of fluid contact is not universally accepted. For example, water-oil contact may be defined as the depth at which the capillary pressure is zero (the free water level). The WOC depth can be identified using a Repeat Formation Test by finding the point of intersection between the oil phase pressure and the water phase pressure. By contrast, water-oil contact may be defined as the deepest point in the reservoir at which a well can still produce water-free oil. The different definitions of contact result in differences in the transition zone model, so it is important to know which definition is applicable and who has the authority to judge the validity of the model. In some cases, it may be necessary to prepare models with both definitions and treat one definition as the base case while the other definition is viewed as a sensitivity.

The proper way to include capillary pressure in a model study is to correct laboratory measured values to reservoir conditions. This is done by applying the correction:

$$P_{c(res)} = P_{c(lab)} \eta_{corr}, \quad \eta_{corr} \equiv \frac{(\gamma \mid \cos \theta \mid)_{res}}{(\gamma \mid \cos \theta \mid)_{lab}}$$

where γ is interfacial tension (IFT) is wettability angle [Amyx, et al., 1960]. The problem with the correction is that it requires data that are seldom known with any certainty, namely interfacial tension and wettability contact angle at reservoir conditions. Alternative approaches include adjusting capillary pressure curves to be consistent with well log estimates of transition zone thickness, or assuming the contact angle factors out. In the latter case, $\eta_{corr} \approx \gamma_{res}/\gamma_{lab}$ [Amyx, et al., 1960]. If laboratory measurements of IFT are not available, IFT can be estimated from the Macleod-Sugden correlation for pure compounds or the Weinaug-Katz correlation for mixtures [Fanchi, 1990].

6.3 Viscous Fingering

Viscous fingering is the unstable displacement of a more viscous fluid by a less viscous fluid. The fingering of an injection fluid into an *in situ* fluid can influence reservoir flow behavior and adversely impact recovery. It is important to note, however, that fingering occurs even in the absence of a porous medium. If a low viscosity fluid is injected into a cell containing a high viscosity fluid, the low viscosity fluid will begin to form fingers as it moves through the fluid. It will not uniformly displace the higher viscosity fluid. These fingers can have different shapes. Figure 6-6 shows an example of a "skeletal" finger [Daccord, et al., 1986] while Figure 6-7 illustrates "fleshy" fingers [for example, see Paterson, 1985; Fanchi and Christiansen, 1989]. If we watch fingers evolve in a homogeneous medium (Figure 6-7), we see fingering display a symmetric pattern. The symmetry can be lost if there is some heterogeneity in the system.

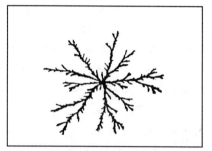

Figure 6-6. "Skeletal" viscous finger (after Daccord, et al. 1986; reprinted by permission of the American Physical Society).

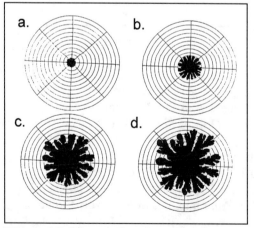

Figure 6-7. Viscous fingering (Fanchi and Christiansen, 1989; reprinted by permission of the Society of Petroleum Engineers).

Fingering can be a reservoir heterogeneity problem or a fluid displacement problem. Most reservoir simulators do not accurately model fingering effects. It is possible to improve model accuracy by using a very fine grid to cover the area of interest, but the benefits associated with such a fine grid are seldom sufficient to justify the additional cost.

Exercises

Exercise 6.1 Data set EXAM3.DAT is a model of a Buckley-Leverett waterflood. What effect does multiplying horizontal permeability by 0.1 and 10 have on EXAM3.DAT results? Consider breakthrough times (time when water production begins), water-oil ratio, and cumulative oil produced at the end of the run. See Chapters 19.3.1 and 19.3.2 for a description of permeability input data. Cumulative production can be found in BTEMP.PLT.

Exercise 6.2 Repeat Exercise 6.1, but modify horizontal transmissibility instead of horizontal permeability. See Chapter 19.3.3. for details.

Exercise 6.3 Determine the effect of doubling water relative permeability in EXAM3.DAT.on water-oil ratio and breakthrough times.

References

Amyx, J.W., D.H. Bass, and R.L. Whiting (1960): **Petroleum Reservoir Engineering**, New York: McGraw-Hill.

Bear, J. (1972): **Dynamics of Fluids in Porous Media**, New York: Elsevier.

Collins, R.E. (1961): **Flow of Fluids Through Porous Materials**, Tulsa, OK: PennWell Publishing.

Daccord, G., J. Nittmann and H.E. Stanley (1986): "Radial Viscous Fingers and Diffusion-Limited Aggregation: Fractal Dimension and Growth Sites," *Physical Review Letters*, Volume 56, pp. 336-339.

Dake, L.P. (1978): **Fundamentals of Reservoir Engineering**, Amsterdam: Elsevier.

Fanchi, J.R. (1983): "Multidimensional Numerical Dispersion," *Society of Petroleum Engineers Journal*, pp. 143-151.

Fanchi, J.R. (1990): "Calculation of Parachors for Compositional Simulation: An Update," *Society of Petroleum Engineers Reservoir Engineering*, pp. 433-436.

Fanchi, J.R. and R.L. Christiansen (1989): "Applicability of Fractals to the Description of Viscous Fingering," Paper SPE 19782, *Proceedings of 64th Annual Technology Conference And Exhibit of Society of Petroleum Engineers*, San Antonio, TX, Oct. 8-11.

Fanchi, J.R., H.-Z. Meng, R.P. Stoltz, and M.W. Owen (1996): "Nash Reservoir Management Study with Stochastic Images: A Case Study," *Society of Petroleum Engineers Formation Evaluation*, pp. 155-161.

Govier, G.W., Editor (1978): **Theory and Practice of the Testing of Gas Wells**, Calgary: Energy Resources Conservation Board.

Honarpour, M., L.F. Koederitz, and A.H. Harvey (1982): "Empirical Equations for Estimating Two-Phase Relative Permeability in Consolidated Rock," *Journal of Petroleum Technology*, pp. 2905-2908.

Kasap, E. and L.W. Lake (June 1990): "Calculating the Effective Permeability Tensor of a Gridblock," *Society of Petroleum Engineers Formation Evaluation*, pp. 192-200.

Koederitz, L.F., A.H. Harvey, and M. Honarpour (1989): **Introduction to Petroleum Reservoir Analysis**, Houston: Gulf Publishing.

Lake, L.W. (April 1988): "The Origins of Anisotropy," *Journal of Petroleum Technology*, pp. 395-396.

Mattax, C.C. and R.L. Dalton (1990): **Reservoir Simulation**, SPE Monograph #13, Richardson, TX: Society of Petroleum Engineers.

Paterson, L. (January 1985): "Fingering with Miscible Fluids in a Hele-Shaw Cell," *Physics of Fluids,* Volume 28 (1), pp. 26-30.

Schneider, F.N. (May 1987): "Three Procedures Enhance Relative Permeability Data," *Oil & Gas Journal*, pp. 45-51.

Chapter 7

Fundamentals of Reservoir Simulation

Previous chapters describe much of the data that is needed by a reservoir simulator. Our goal here is to outline the physical, mathematical and computational basis of reservoir flow simulation. For a more detailed technical presentation, consult one of the many sources available in the literature [for example, see Aziz and Settari, 1979; Bear, 1972; Mattax and Dalton, 1990; Peaceman, 1977; and Thomas, 1982]. The set of equations used in BOAST4D is derived in Chapter 32.

7.1 Conservation Laws

The basic conservation laws of reservoir simulation are the conservation of mass, energy and momentum. Mass balance in a representative elementary volume (REV) or grid block is achieved by equating the accumulation of mass in the block with the difference between the mass leaving the block and the mass entering the block. The set of equations used in BOAST4D are derived from the mass conservation principle in Chapter 32. A material balance is performed for each block. What makes a simulator different from a reservoir engineering material balance program is the ability of the simulator to account for flow between blocks.

A material balance calculation is actually a subset of the simulator capability. This is an important point because it means a reservoir simulator can be used to perform material balance work. The advantage of using a simulator

58

instead of a material balance program is that the simulation model can be enlarged to include position-dependent effects by modifying the grid representing the reservoir architecture. Thus, a single block material balance calculation in a reservoir simulation model can be expanded with relative ease to include flow in one, two or three spatial dimensions. This procedure is used in the case study presented in Part II.

Most reservoir simulators assume isothermal conditions. They also assume complete and instantaneous phase equilibration in each node. Thus, most simulators do not account for the time it takes a mixture to reach equilibrium; they assume, instead, that equilibration is established instantaneously. This is often a reasonable assumption.

Momentum conservation is modeled using Darcy's Law. This assumption means that the model does not accurately represent turbulent flow in a reservoir or near the wellbore. Some well models allow the user to model turbulent flow, especially for high flow rate gas wells. Turbulent flow models relate pressure change to a linear flow term, as in Darcy's Law, plus a term that is quadratic in flow rate. This quadratic effect is not usually included in the reservoir model, only in the well model.

7.2 Flow Equations

The general equations for describing fluid flow in a porous medium are shown in Table 7-1 and associated nomenclature is presented in Table 7-2. The molar conservation equation includes a dispersion term, a convection term, a source/sink term representing wells, and the time varying accumulation term. The dispersion term is usually neglected in most workhorse simulators such as black oil and compositional simulators (see Chapters 23 and 32). Neglecting dispersion simplifies program coding and is justified when dispersion is a second-order effect. In some situations, such as miscible gas injection, physical dispersion is an effect that should be considered. Further discussion of dispersion is presented in Chapter 8.

Table 7-1
Molar Conservation Equation for Component k

Physical Source	Term
Dispersion	$\nabla \cdot \left[\sum_{\ell=1}^{n_p} \phi \, S_\ell \, \underline{\underline{D}}_{k\ell} \, \rho_\ell \cdot \nabla x_{k\ell} \right]$
Convection	$-\nabla \cdot \left[\sum_{\ell=1}^{n_p} \rho_\ell \, x_{k\ell} \, V_\ell \right]$
Source/Sink	$+ \, Q_k$
Accumulation	$= \dfrac{\partial}{\partial t} \left[\phi \sum_{\ell=1}^{n_p} \rho_\ell \, x_{k\ell} \, S_\ell \right]$
Darcy's Law	$V_\ell = -\underline{\underline{K}} \dfrac{k_{r\ell}}{\mu_\ell} \cdot (\nabla P_\ell - \gamma_\ell \nabla z)$

Table 7-2
Terminology of Molar Conservation Equation

Variable	Meaning
$\underline{\underline{D}}_{kl}$	Dispersion tensor of component k in phase ℓ
$\underline{\underline{K}}$	Permeability tensor
k_{rl}	Relative permeability of phase ℓ
n_c	Number of components
n_p	Number of phases
P_ℓ	Pressure of phase ℓ
S_ℓ	Saturation of phase ℓ
V_ℓ	Darcy's velocity for phase ℓ
x_{kl}	Mole fraction of component k in phase ℓ
γ_ℓ	Pressure gradient of phase ℓ
μ_ℓ	Viscosity of phase ℓ
ρ_ℓ	Density of phase ℓ
ϕ	Porosity

The molar flow equations were derived using the conservation laws introduced in Chapter 7.1. An energy balance equation can be found in the thermal recovery literature [Prats, 1982]. The energy balance equation is more complex than the flow equations because of the presence of additional nonlinear terms. Energy loss to adjacent non-reservoir rock must also be computed. The resulting complexity requires substantial computation to achieve an energy balance. In many realistic systems, reservoir temperature variation is slight and the energy balance equation can be neglected by imposing the isothermal approximation. The result is a substantial savings in computation expense with a reasonably small loss of accuracy.

Several supplemental – or auxiliary – equations must be specified to complete the definition of the mathematical problem. There must be a flow equation for each modeled phase. Commercial black oil and compositional simulators are formulated to model up to three phases: oil, water, and gas. The inclusion of gas in the water phase can be found in some simulators, though it is neglected in most. The ability to model gas solubility in water is useful for CO_2 floods or for modeling geopressured gas-water reservoirs. Some black oil simulator formulations include a condensate term. It accounts for liquid yield associated with condensate reservoir performance.

In addition to modeling reservoir structure and PVT data, simulators must include rate equations for modeling wells, phase potential calculations, and rock-fluid interaction data such as relative permeability curves and capillary pressure curves. Saturation dependent rock-fluid interaction data are entered in either tabular or analytical form. More sophisticated simulators let the user represent different types of saturation change processes, such as imbibition, drainage, and hysteresis. Applying such options leads to additional computation and cost.

7.3 Well and Facilities Modeling

Well and surface facility models are simplified representations of real equipment [Williamson and Chappelear, 1981]. The well model, for example, does not account for flow in the wellbore from the reservoir to the surface. This effect can be taken into account by adding a wellbore model. The wellbore model usually consists of a multivariable table relating surface pressure to such parameters as flow rate and GOR. The tables are often calculated using a separate program that performs a nodal analysis of wellbore flow. Well models typically assume that fluid phases are fully dispersed and that the block containing the well is perforated throughout its thickness. Some commercial simulators will let the user specify a perforation interval under certain conditions.

The different types of well controls include production and injection well controls, and group and field controls for a surface model. The production well model assumes the user specifies one option as the primary control, but may also specify other options as targets for constraining the primary control. For example, if oil rate is the primary control, then the produced GOR may be restricted so that the oil rate is decreased when GOR exceeds the specified value. This provides a more realistic representation of actual field practice.

Injection well controls assume that initial injection well mobility is given by total grid block mobility. This makes it possible to inject a phase into a block that would otherwise have zero relative permeability to flow.

Allocation of fluids in a well model depends on layer flow capacity and fluid mobility. The fluid allocation procedure in BOAST4D is discussed in Chapter 28. Simulators can also describe deviated or horizontal wells depending on how well completions and parameters are specified.

Well, group and field controls can be specified in commercial simulators with a surface facilities model. The user specifies a hierarchy of controls that most realistically represent how the field is being operated. For example, well production may be constrained by platform separator and storage capacity, which in turn is constrained by pipeline flow capacity.

7.4 Simulator Solution Procedures

Fluid flow equations are a set of nonlinear partial differential equations that must be solved by computer. The partial derivatives are replaced with finite differences, which are in turn derived from Taylor's series [for example, see Aziz and Settari, 1979; Peaceman, 1977; Rosenberg, 1977; Fanchi, 1997]. This procedure is illustrated in Table 7-3. The spatial finite difference interval Δx

along the x-axis is called grid block length, and the temporal finite difference interval Δt is called the time step. Indices i, j, k are ordinarily used to label grid locations along the x, y, z coordinate axes, respectively. Index n labels the present time level, so that $n + 1$ represents a future time level. If the finite difference representations of the partial derivatives are substituted into the original flow equations, the result is a set of equations that can be algebraically rearranged to form a set of equations that can be solved numerically. The solution of these equations is the job of the simulator.

Table 7-3

Finite Difference Approximation

♦ Formulate fluid flow equations, such as,

$$\frac{\partial}{\partial x}\left[\frac{Kk_r}{\mu B}\left(\frac{\partial P}{\partial x}\right)\right] + q_s \delta(x - x_0) = \frac{\partial}{\partial t}\left(\frac{\phi S}{B}\right)$$

♦ Approximate derivatives with finite differences

◊ Discretize region into grid blocks Δx:

$$\frac{\partial P}{\partial x} \approx \frac{P_{i+1} - P_i}{x_{i+1} - x_i} \equiv \frac{\Delta P}{\Delta x}$$

◊ Discretize time into time steps Δt:

$$\frac{\partial S}{\partial t} \approx \frac{S^{n+1} - S^n}{t^{n+1} - t^n} \equiv \frac{\Delta S}{\Delta t}$$

♦ Numerically solve the resulting set of linear algebraic equations

---

The two most common solution procedures in use today are IMPES and Newton-Raphson. The terms in the finite difference form of the flow equations are expanded in the Newton-Raphson procedure as the sum of each term at the current iteration level, plus a contribution due to a change of each term with respect to the primary unknown variables over the iteration. To calculate these changes, it is necessary to calculate derivatives, either numerically or analytically, of the flow equation terms. The derivatives are stored in a matrix called the acceleration matrix or the Jacobian. The Newton-Raphson technique leads to a matrix equation $J \cdot \delta X = R$ that equates the product of the acceleration matrix J and a column vector δX of changes to the primary unknown variables to the column vector of residuals R. It is solved by matrix algebra to yield the changes to the primary unknown variables δX. These changes are added to the value of the primary unknown variables at the beginning of the iteration. If the changes are less than a specified tolerance, the iterative Newton-Raphson technique is considered complete and the simulator proceeds to the next time step.

The three primary unknown variables for an oil-water-gas system are oil phase pressure, water saturation, and either gas saturation or solution GOR. The choice of the third variable depends on whether or not the block contains free gas, which depends, in turn, on whether the block pressure is above or below bubble point pressure. Naturally, the choice of unknowns is different for a gas-water or water only system. This discussion applies to the most general three-phase case.

A simpler procedure is the IMplicit Pressure-Explicit Saturation (IMPES) procedure. It is much like the Newton-Raphson technique except that flow coefficients are not updated in an iterative process. The Newton-Raphson technique is known as a fully implicit technique because all primary variables are calculated at the same time, that is, primary variables at the new time level are determined simultaneously. By contrast, the IMPES procedure solves for pressure at the new time level using saturations at the old time level, then uses the pressures at the new time level to explicitly calculate saturations at the new time level. BOAST4D, the program provided with this book, is an implementation of a noniterative IMPES formulation [Fanchi, et al., 1982; Fanchi, et al., 1987]. The formulation is outlined in Chapters 23 and 33. A variation of this technique is to iteratively substitute the new time level estimates of primary variables in the calculation of coefficients for the flow equations. The iterative IMPES technique takes longer to run than the non-iterative technique, but generates less material balance error [Ammer and Brummert, 1991].

A flow chart for a typical simulator is shown in Figure 7-1 (see Crichlow, 1977). The simulation program begins by reading input data and initializing the reservoir. This part of the model will not change as a function of time. Information for time-dependent data must then be read. This data includes well and field control data. The coefficients of the flow equations and the primary unknown variables are then calculated. Once the primary variables are determined, the process can be repeated by updating the flow coefficients using the values of the primary variables at the new iteration level. This iterative process can improve material balance. When the solution of the flow equations is complete, flow properties are updated and output files are created before the next time step begins.

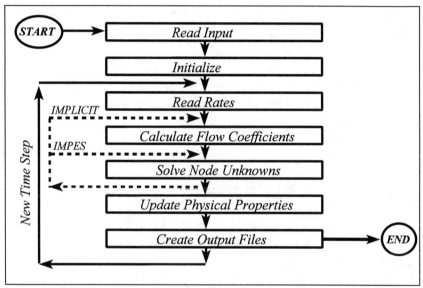

Figure 7-1. Typical simulator flow chart.

Fully implicit techniques do more calculations in a time step than the IMPES procedure, but are stable over longer time steps. The unconditional stability of the fully implicit techniques means that a fully implicit simulator can solve problems faster than IMPES techniques by taking significantly longer time steps.

A problem with large time steps in the fully implicit technique is the introduction of a numerical effect known as numerical dispersion [Lantz, 1971; Fanchi, 1983]. Numerical dispersion is introduced when the Taylor series approximation is used to replace derivatives with finite differences. The resulting truncation error introduces an error in calculating the movement of saturation fronts that looks like physical dispersion, hence it is called numerical dispersion.

Numerical dispersion arises from time and space discretizations that lead to smeared spatial gradients of saturation or concentration [Lantz, 1971] and grid orientation effects [Fanchi, 1983; and Chapter 8]. The smearing of saturation fronts can impact the modeling of displacement processes. An illustration of front smearing is presented in Figure 7-2 for a linear Buckley-Leverett water-flood model. The numerical front from an IMPES calculation does not exhibit the same piston-like displacement that is shown by the analytical Buckley-Leverett calculation [for example, see Collins, 1961; Mian, 1992].

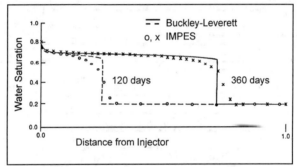

Figure 7-2. Numerical dispersion (after Fanchi, 1986; reprinted by permission of the Society of Petroleum Engineers).

Numerical dispersion D^{num} in one dimension has the form

$$D^{num} = \frac{v}{2}\left(\Delta x \pm \frac{v\Delta t}{\phi}\right)$$

It depends on grid block size Δx, time step size Δt, velocity v of frontal advance, porosity ϕ, and numerical formulation. The "+" sign applies to the fully implicit formulation, and the "–" sign applies to IMPES. Notice that an increase in Δt in the fully implicit formulation increases D^{num} while it decreases D^{num} when the IMPES technique is used. Indeed, it appears that a judicious choice of Δx and Δt could eliminate D^{num} altogether in the IMPES method. Unfortunately, the combination of Δx and Δt that yields $D^{num} = 0$ violates a numerical stability criterion. In general, IMPES numerical dispersion is not as large as that associated with fully implicit techniques.

As a rule of thumb, time step sizes in fully implicit calculations should not exceed a quarter of a year, otherwise numerical dispersion can dominate front modeling. By contrast, the maximum time step size in an IMPES simulator can be estimated by applying the rule of thumb that throughput in any block should not exceed 10% of the pore volume of the block. Throughput is the volume of fluid that passes through a block in a single time step. IMPES time step sizes are often on the order of a month or less. An example of a throughput calculation is given in Chapter 16.4.

The IMPES time step limitation is less of a problem than it might otherwise seem, because it is very common to have production data reported on a monthly basis. The reporting period often controls the frequency with which well control data is read during a history match. Thus, during the history match phase of a study, simulator time step sizes are dictated by the need to enter historical data. Large time step sizes reduce the ability of the model to track variations of rate with time because historical data must be averaged over a longer period of time. As a result, the modeler often has to constrain the fully implicit simulator to run at less than optimum numerical efficiency because of the need to more accurately represent the real behavior of the physical system.

Fully implicit techniques represent the most advanced simulation technology, yet IMPES retains vitality as a relatively inexpensive means of modeling some problems. Unless a fully implicit model is readily available, it is not always necessary nor cost-effective to employ the most advanced technology to solve every reservoir simulation problem. The wise modeler will recognize that you do not have to use a sledge hammer to open a peanut!

Simulators also differ in their robustness, that is, their ability to solve a wide range of physically distinct problems. Robustness appears to depend as much on the coding of the simulator as it does on the formulation technique. The best way to determine simulator robustness is to test the simulator with data sets representing many different types of reservoir management problems. The examples provided with BOAST4D are designed to demonstrate the robustness, or range of applicability, of the simulator.

Simulator technology is generally considered proprietary technology, yet it has an economic impact that takes it out of the realm of the research laboratory and makes it a topic of importance in the corporate boardroom. Nevertheless, numerical representations of nature are subject to inaccuracies [for example, see Mattax and Dalton, 1990; Saleri, 1993; and Oreskes, et al., 1994]. This point has been illustrated in several simulator comparison projects sponsored by the Society of Petroleum Engineers beginning with Odeh [1981] and continuing through Killough [1995]. Each comparison project was designed to allow comparisons of proprietary technology by asking participating organizations to solve the same pre-determined problem. Figure 7-3 is taken from the first comparison project [Odeh, 1981]. The first project compared the performance of simulators modeling the injection of gas into a saturated black oil reservoir. Figure 7-3 shows that differences in the formulations of several reservoir simulators lead to differences in predictions of economically important quantities such as oil rate production.

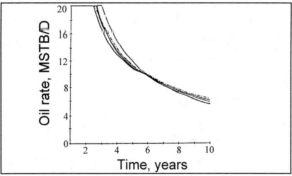

Figure 7-3. Oil rate from first SPE comparative solution project (after Odeh, 1981; reprinted by permission of the Society of Petroleum Engineers).

In summary, a representation of the reservoir is quantified in the simulator. The representation is validated during the history matching process, and forecasts of reservoir performance are then made from the validated reservoir representation.

7.5 Simulator Selection

The selection of a reservoir simulator depends on such factors as the objectives of the study, fluid type, and dimensionality of the system. For purposes of illustration, we focus our attention on a study which uses either a black oil or a compositional simulator. Standard black oil and compositional simulators assume isothermal flow and mass transfer within a block is instantaneous. A compositional simulator represents the fluid as a mixture of hydrocarbon components. Black oil simulators may be viewed as compositional simulators with two components. They can have gas dissolved in the oil phase, as well as oil dissolved in the gas phase. Black oil simulators need both saturated and undersaturated fluid property data, as discussed in Chapter 5.

Black oil and compositional simulators usually assume fluids have a minimal effect on rock properties. Thus, standard versions of the simulators will not model changes in rock properties due to effects like grain dissolution, tar mat formation, or gel formation resulting from a vertical conformance treatment. Special purpose simulators or special options within a standard simulator must be obtained to solve such problems.

Fluid type is needed to decide if the reservoir should be modeled using either a black oil simulator or a compositional simulator. Well logs can distinguish between oil and gas, but are less useful in further classifying fluid type. A pressure-temperature diagram is useful for determining reservoir fluid type, but its preparation requires laboratory work with a fluid sample. A simpler way that is often sufficient for classifying a fluid is to look at solution gas-oil ratio. Table 5-1 shows typical solution GOR ranges for each fluid type. As a rule of thumb, compositional models should be used to model volatile oil and condensate fluids, while black oil and dry gas fluids are most effectively modeled with a black oil simulator. The applicability of this rule depends on the objectives of the study.

The pressure range associated with fluid property data should cover the entire range of pressures expected to be encountered over the life of the field. The data should be smooth to enhance computational efficiency and to ensure data consistency. A check on data consistency is a calculation of fluid compressibility. If a negative compressibility is encountered, the data need to be corrected. The problem of negative compressibility occurs most often when data is extrapolated beyond measured pressure ranges.

Flow units should be determined by reviewing geological and petrophysical data. It is possible to represent the behavior of a flow unit by defining a set of PVT and Rock property tables for each flow unit. PVT property tables contain data that describe fluid properties, while Rock property tables represent relative permeability and capillary pressure effects. Each set of PVT or Rock property tables applies to a particular region of grid blocks, hence the collection of blocks to which a particular set of PVT or Rock property tables applies is referred to as a PVT or Rock region. The number of flow units, and the corresponding number of PVT and Rock regions, should be kept to the minimum needed to achieve the objectives of the study. This statement is another application of Ockham's Razor (Chapter 9.3).

Exercises

Exercise 7.1 Data file EXAM8.DAT has a gas well under LIT control Determine the effect of doubling the turbulence factor. See Chapters 20.2 and 28.2 for more discussion.

Exercise 7.2 BOAST4D contains a few fieldwide controls (see Chapter 19.8). Data file EXAM4.DAT is a 2D areal model of an undersaturated oil reservoir undergoing primary depletion. Modify data file EXAM4.DAT so that fieldwide pressure is not allowed to drop below the initial bubble point pressure. The initial bubble point pressure is also described in Chapter 19.6. What effect does this have on the duration of the run?

Exercise 7.3 Data set EXAM3.DAT can be used to study the numerical dispersion associated with a Buckley-Leverett type waterflood of an undersaturated oil reservoir. Run EXAM3.DAT with constant time steps of 5 days, 10 days, and 15 days. Rerun the problem with time step size beginning at 5 days and allowed to vary from 5 days to 15 days. How does the water breakthrough time (time when the model reaches a water-oil ratio of 0.1) change from one case to another? Time step controls are discussed in Chapters 19.8 and 20.1.

Exercise 7.4 Data set EXAM7.DAT is one version of the Odeh [1981] SPE comparative solution problem. Run EXAM7.DAT and compare the results to those reported by Odeh. What is the BOAST4D material balance error? The material balance error associated with this data set provides a good test of the quality of BOAST4D relative to other programs based on the original version of BOAST [for example, Fanchi, et al., 1982; Fanchi, et al., 1987; Louisiana State University, 1997].

Exercise 7.5 Data set EXAM10.DAT illustrates the use of PVT and Rock regions in BOAST4D (see Chapter 19.4). Run EXAM10.DAT and determine the number of regions in the data set.

References

Ammer, J.R. and A.C. Brummert (1991): "Miscible Applied Simulation Techniques for Energy Recovery – Version 2.0," U.S. Department of Energy Report DOE/BC-91/2/SP, Morgantown Energy Technology Center, WV.

Aziz, K. and A. Settari (1979): **Petroleum Reservoir Simulation**, New York: Elsevier.

Bear, J. (1972): **Dynamics of Fluids in Porous Media**, New York: Elsevier.

Collins, R.E. (1961): **Flow of Fluids Through Porous Materials**, Tulsa, OK: PennWell Publishing.

Fanchi, J.R. (1983): "Multidimensional Numerical Dispersion," *Society of Petroleum Engineers Journal.*, pp. 143-151.

Fanchi, J.R. (1986): "BOAST-DRC: Black Oil and Condensate Reservoir Simulation on an IBM-PC," Paper SPE 15297, *Proceedings of the Symposium On Petroleum Industry Applications of Microcomputers of the Society of Petroleum Engineers*, Silver Creek, CO, June 18-20.

Fanchi, J.R. (1997): **Math Refresher for Scientists and Engineers**, New York: J. Wiley & Sons.

Fanchi, J.R., K.J. Harpole, and S.W. Bujnowski (1982): "BOAST: A Three-Dimensional, Three-Phase Black Oil Applied Simulation Tool", 2 Volumes, U.S. Department of Energy, Bartlesville Energy Technology Center, OK.

Fanchi, J.R., J. E. Kennedy, and D.L. Dauben (1987): "BOAST II: A Three-Dimensional, Three-Phase Black Oil Applied Simulation Tool," U.S. Department of Energy, Bartlesville Energy Technology Center, OK.

Killough, J.E. (1995): "Ninth SPE Comparative Solution Project: A Reexamination of Black-Oil Simulation," Paper SPE 29110, *Proceedings of the 13th Society of Petroleum Engineers Symposium On Reservoir Simulation*, Feb. 12-15.

Lantz, R.B. (1971): "Quantitative Evaluation of Numerical Diffusion," *Society of Petroleum Engineering Journal*, pp. 315-320.

Louisiana State University (1997): " 'BOAST 3' A Modified Version of BOAST II with Post Processors B3PLOT2 and COLORGRID," Version 1.50, U.S. Department of Energy Report DOE/BC/14831-18, Bartlesville Energy Technology Center, OK.

Mattax, C.C. and R.L. Dalton (1990): **Reservoir Simulation**, SPE Monograph #13, Richardson, TX: Society of Petroleum Engineers.

Mian, M.A. (1992): **Petroleum Engineering Handbook for the Practicing Engineer**, Volumes I and II, Tulsa, OK: PennWell Publishing.

Odeh, A.S. (1981): "Comparison of Solutions to a Three Dimensional Black-Oil Reservoir Simulation Problem," *Journal of Petroleum Technology*, pp. 13-25.

Oreskes, N., K. Shrader-Frechette, and K. Belitz (1994): "Verification, Validation, and Confirmation of Numerical Models in the Earth Sciences", *Science*, pp. 641-646, Feb. 4.

Peaceman, D.W. (1977): **Fundamentals of Numerical Reservoir Simulation**, New York: Elsevier.

Prats, M. (1982): **Thermal Recovery**, SPE Monograph Series, Richardson TX: Society of Petroleum Engineers.

Rosenberg, D. U. von (1977): **Methods for the Numerical Solution of Partial Differential Equations**, Tulsa, OK: Farrar and Associates.

Saleri, N.G. (1993): "Reservoir Performance Forecasting: Acceleration by Parallel Planning," *Journal of Petroleum Technology*, pp. 652-657.

Thomas, G.W. (1982): **Principles of Hydrocarbon Reservoir Simulation**, Boston: International Human Resources Development Corporation.

Williamson, A.E. and J.E. Chappelear (June 1981): "Representing Wells in Numerical Reservoir Simulation: Part I – Theory," *Society of Petroleum Engineering Journal*, pp. 323-338; and "Part II – Implementation,"*Society of Petroleum Engineering Journal*, pp. 339-344.

Chapter 8

Modeling Reservoir Architecture

Reservoir architecture is modeled by contouring and digitizing geologic maps. The mapping/contouring process is the point where the geological and geophysical interpretations have their greatest impact on the final representation of the reservoir. This process has been discussed by several authors, including Harpole [1985], Harris [1987], and Tearpock and Bischke [1991]. Methods for numerically representing reservoir architecture are discussed in this chapter.

8.1 Mapping

The different parameters that must be digitized for use in a grid include elevations or structure tops, permeability in three orthogonal directions, porosity, gross thickness, net to gross thickness and, where appropriate, descriptions of faults, fractures and aquifers. The resulting maps are digitized by overlaying a grid on the maps and reading a value for each grid block. The digitizing process is sketched in Figures 8-1a through 8-1d.

The resolution of the model depends on the resolution of the grid. A fine grid divides the reservoir into many small grid blocks. It gives the most accurate numerical representation, but has the greatest computational expense. A coarse grid has fewer grid blocks, but the coarse grid blocks must be larger than the fine grid blocks to cover the same model volume. As a result, the coarse grid is less expensive to run than a fine grid, but it is also less accurate numerically.

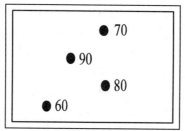

Figure 8-1a. Gather data.

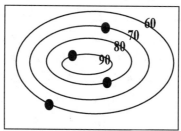

Figure 8-1b. Contour data.

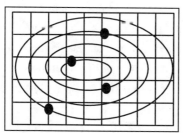

Figure 8-1c. Overlay grid.

60	60	60	65	65	65	60	60	60
60	60	75	80	82	80	75	67	60
65	75	85	90	90	86	80	70	64
60	70	75	77	78	77	74	65	60
60	60	60	65	66	65	62	60	60

Figure 8-1d. Digitize data.

The loss of accuracy is most evident when a coarse grid is used to model the interfaces between phases such as fluid contacts and displacement fronts. Thus, fine grid modeling is often the preferred choice to achieve maximum numerical accuracy. It is important to recognize, however, that a fine grid covering an area defined by sparse data can give the illusion of accuracy. Sensitivity studies can help quantify the uncertainty associated with the model study.

The gridding process is most versatile when used with an integrated 3D reservoir mapping package. Modern mapping techniques include computer generated maps that can be changed relatively quickly once properly set up. Dahlberg [1975] presented one of the first analyses of the relative merits of hand drawn and computer generated maps. Computer generated maps may not include all of the detailed interpretation a geologist might wish to include in the model, particularly with regard to faults, but the maps generated by computer in a 3D mapping program do not have the problem so often associated with the stacking of 2D plan view maps, namely physically unrealistic layer overlaps. Layer overlaps need to be corrected before the history match process begins.

Another problem with computer generated maps is the amount of detail that can be obtained. Computer generated maps can describe a reservoir with a much finer grid than can be used in a reservoir simulator. For example, a computer mapping program such as that described by Englund and Sparks [1991] or Pannatier [1996] may use a grid with a million or more cells to represent the reservoir, yet reservoir simulation grids are usually 100,000 blocks or less. This means that the reservoir representation in the computer mapping program must be scaled up, or coarsened, for use in a reservoir simulator. Although many attempts have been made to find the most realistic process for scaling up data, there is no widely accepted method in use today [for example, see Christie, 1996].

8.2 Grid Preparation

Reservoir grids may be designed in several different ways. For a review of different types of grids, see Aziz [1993]. Definitions of coordinate system orientation vary from one simulator to another and must be clearly defined for effective use in a simulator. Reservoir grids can often be constructed in one-, two-, or three-dimensions, and in Cartesian or cylindrical coordinates. Horizontal 1D models are used to model linear systems that do not include gravity effects. Examples of horizontal 1D models include core floods and linear displacement in a horizontal layer. Core flood modeling has a variety of applications, including the determination of saturation dependent data such as relative permeability curves. A dipping 1D reservoir is easily defined in a model by specifying structure top as a function of distance from the origin of a grid.

Figure 8-1 is an example of a 2D grid. Grids in 2D may be used to model areal and cross-sectional fluid movement. Grid orientation in 2D is illustrated by comparing Figure 8-1c and Figure 8-2. Although Figure 8-1c has fewer blocks, which is computationally more efficient, Figure 8-2 may be useful in some circumstances. For example, Figure 8-2 is more useful than Figure 8-1c if the boundary of the reservoir is not well known or an aquifer needs to be attached to the flanks of the reservoir to match reservoir behavior.

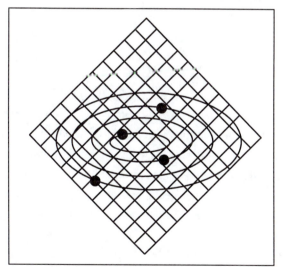

Figure 8-2. Grid orientation.

The use of 2D grids for full field modeling has continued to be popular even as computer power has increased and made large 3D models practical. Figure 8-3 shows a simple 3D grid that is often called a "layer cake" grid. Techniques are available for approximating the vertical distribution of fluids in 2D cross-sectional and 3D models by modifying relative permeability and capillary pressure curves. The modified curves are called pseudo curves. Taggart, et al. [1995] present a discussion of several pseudoization techniques and their limitations. An example of a pseudoization technique is the vertical equilibrium (VE) assumption. The principal VE assumption is that fluid segregation in the vertical dimension is instantaneous. This assumption is approximated in nature when vertical flow is rapid, as in the case when the reservoir exhibits large vertical permeability and when density differences are significant, such as in gas-oil or gas-water systems. For more discussion of specific pseudoization techniques, see Taggart, et al. [1995] and their references.

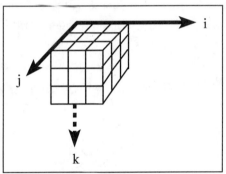

Figure 8-3. Example of a 3D "layer cake" grid.

One reason for the continuing popularity of 2D grids is that the expectation of what is appropriate grid resolution has changed as simulation technology evolved. Thus, even though 3D models could be used today with the grid resolution that was considered acceptable a decade ago for 2D models, modern expectations often require that even finer grids be used for the same types of problems. This is an example of a task expanding to fit the available resources. It is not obvious that the increased grid definition is leading to better reservoir management decisions. Indeed, it can be argued that the technological ability to add complexity is making it more difficult for people to develop a "big picture" understanding of the system being studied because they are too busy focusing on the details of a complex model. Once again, a judicious use of

Ockham's Razor is advisable in selecting a reservoir grid. The grid should be appropriate for achieving study objectives.

Near wellbore coning models may be either 2D or 3D grids, but are defined in cylindrical rather than Cartesian coordinates. Coning (or radial) models are designed to study rapid pressure and saturation changes. An example of a radial grid is shown in Figure 8-8. High throughput, that is, large flow rate, in relatively small, near wellbore grid blocks is most effectively simulated by a fully implicit formulation. IMPES can be used to model coning, but time steps must be very small, possibly on the order of minutes or hours. Small time steps are not a problem if the duration of the modeled history is short, as it would be in the case of a pressure transient test.

Grid blocks may be defined in terms of corner point geometry or block centered geometry (Figure 8-4). Block centered geometry is the most straightforward technique, but corner point geometry has gained popularity because it yields more visually realistic representations of reservoir architecture. This is valuable when making presentations to nonspecialists. The different geometric representations are illustrated for a two-layer dipping reservoir in Figure 8-5. Although corner point geometry is visually more realistic, it is easier to define a grid with block centered geometry. Block centered geometry simply requires the specification of the lengths of each side of the block and the block center or top. Corner point geometry requires specifying the location of all eight corners of the block. This is most readily accomplished with a computer program.

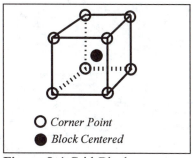

O *Corner Point*
● *Block Centered*

Figure 8-4 Grid Block Representation

There is little computational difference between the results of corner point and block centered geometry. One caution should be noted with respect to corner point geometry. It is possible to define very irregularly shaped grids using corner points. This can lead to the distortion of flood fronts and numerical stability problems if the irregularities from a parallelogram are severe. Flood front distortion caused by gridding is an example of the grid orientation effect discussed by many authors, including Aziz and Settari [1979], and Mattax and Dalton [1990].

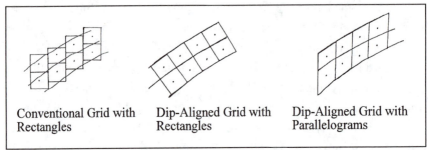

| Conventional Grid with Rectangles | Dip-Aligned Grid with Rectangles | Dip-Aligned Grid with Parallelograms |

Figure 8-5. Geometric representations of a dipping reservoir.

The grid orientation effect is exhibited by looking at a displacement process in 2D (Figure 8-6). Each producer is equidistant from the single injector in a model that has uniform and isotropic properties. If grid orientation did not

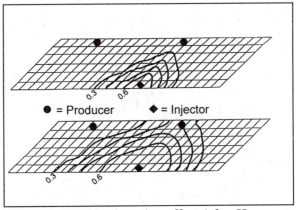

Figure 8-6. Grid orientation effect (after Hegre, et al. 1986; reprinted by permission of the Society of Petroleum Engineers)

orientation did not matter, the symmetry of the problem would show that both wells would produce injected water at the same time. The figure shows that production is not the same. Injected fluids preferentially follow the most direct grid path to the producer. Thus, even though the producers are symmetrically located relative to the injector and each other, the grid orientation altered the expected flow pattern. Figure 8-6 shows the effect on frontal advance.

Another example of the grid orientation effect arises in connection with the modeling of pattern floods. Figure 8-7 illustrates two grids that can be used to model flow in a five-spot pattern. The parallel grid results in earlier breakthrough of injected fluids than the diagonal grid. This effect can be traced to the finite difference representation of the fluid flow equations.

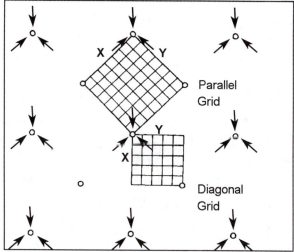

Figure 8-7. Parallel and diagonal grids (after Todd, et al. 1972; reprinted by permission of the Society of Petroleum Engineers)

Most finite difference simulators only account for flow contributions from blocks that are nearest neighbors to the central block along orthogonal Cartesian axes. In Table 8-1, the central block is denoted by "C" and the nearest neighbor blocks contributing to the standard finite difference calculation in 2D are denoted by an asterisk. The five blocks represent the five-point differencing scheme.

Table 8-1

Finite Difference Stencils

Block	I - 1	I	I + 1
J - 1	9	*	9
J	*	C	*
J + 1	9	*	9

Reservoir simulators are usually formulated with the assumption that diagonal blocks do not contribute because the grid is aligned along the principal axes of the permeability tensor. Diagonal blocks are denoted by "9" in Table 8-1. The nine-point stencil includes all nine blocks in the calculation of flow into and out of the central block. Grid orientation effects can be minimized, at least in principle, if the diagonal blocks are included in a nine-point finite difference formulation [for example, see Young, 1984; Hegre, et al., 1986; Lee, et al., 1997]. This option is available in some commercial simulators.

Local grid refinement (LGR) is used to provide additional grid definition in a few selected regions of a larger grid. Raleigh [1991] compared local grid refinement with a radial grid (Figure 8-8) and showed that the results are comparable. When LGR is used, it typically increases computer processor time for a run because of increased throughput in small blocks.

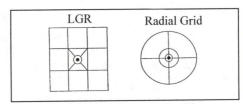

Figure 8-8. LGR and radial grids.

Although many grid preparation options are available, improving grid preparation capability is an ongoing research and development topic. As noted above, grid preparation can introduce unphysical effects into model results. Aziz [1993] and Chin [1993] provide additional discussion of grid preparation research.

8.3 Model Types

Models may be classified into three different types: full field models, window area models, and conceptual models. Full field models are used to match performance of the entire field. They take into account the interaction between all wells and layers. The results of full field models are already matched to field scale and require no further scaling. The disadvantage of using full field models is that the number of grid blocks may need to be large or the grid size may need to be relatively coarse to include the entire field.

Window area models are designed to look at smaller areas of the field. These models are often constructed from a full field description. Window area models allow finer grid resolution or shorter turnaround time if the model runs faster than a full field model. The window area models are useful for studying recovery mechanisms and for determining reasonable grid preparation criteria for use in full field models, especially with regard to layering. Full field models require sufficient layering to track fluid contact movement or other depth dependent information that is needed to achieve study objectives. Window area models have the disadvantage of not being able to accurately model flux across window area boundaries. This means that effects of wells outside the window area are not taken into account except through boundary conditions. Some commercial simulators will output time dependent boundary conditions for use in window area models. Although this information is helpful, the process is cumbersome and does not necessarily yield accurate results. Field history can be used to guide development of the window area model, but has only limited utility as a criterion for validating window model performance.

One of the most useful types of models is the conceptual model. Conceptual models can be built quickly and require only an approximate description of that part of the reservoir that is relevant to the conceptual study. Computer resource requirements are relatively small when compared with full field or window area models. Results of the conceptual model are qualitative and best used for comparing concepts such as vertical layering. They can also be used to prepare pseudo curves for use in full field or window area models. For example, the saturation of a block in a model with a transition zone depends on the depth of the center point of the block (see Chapter 6). As a result, a grid that is vertically coarse may have only a rough approximation of the transition zone. More accurate modeling of saturation gradient in a transition zone requires vertical grid refinement or use of pseudo curves. Conceptual models are useful for preparing such pseudo curves. The disadvantage of conceptual models is that their results do not apply directly to the description of the field. Since there is no history match, conceptual model results should be viewed as qualitative rather than quantitative estimates of field performance.

8.4 Basic Simulator Volumetrics

Reservoir simulators calculate reservoir volume using a procedure similar to the procedure described in this section. Bulk volume V_B of each grid block defined in a Cartesian coordinate system $\{x, y, z\}$ is calculated from the gross thickness $\Delta z = h$ of each grid block and the grid block lengths Δx, Δy along the x and y axes:

$$V_B = h\,\Delta x\,\Delta y$$

Porosity ϕ and net-to-gross ratio η are then used to calculate grid block pore volume

$$V_P = \phi\eta V_B = \phi\eta h\Delta x\Delta y = \phi h_{net}\Delta x\Delta y$$

where net thickness is defined by $h_{net} = \eta\, h$. The volume of phase ℓ in the grid block at reservoir conditions is the product of the grid block pore volume and phase saturation, thus

$$V_\ell = S_\ell V_P = S_\ell\phi h_{net}\Delta x\Delta y$$

where S_ℓ is the saturation of phase ℓ. Total model volumes are calculated by summing over all grid blocks.

Many commercial simulators provide optional variations on the simple procedure outlined above. A comparison of reservoir simulator calculated volumetrics with volumetrics from another source, such as a material balance study or a computer mapping package, provides a means of validating volumetric estimates using independent sources.

Exercises

Exercise 8.1 Sketch the model grids for data sets EXAM1.DAT, EXAM2.DAT, EXAM3.DAT, EXAM5.DAT and EXAM7.DAT using information from each data set.

Exercise 8.2 Repeat Exercise 8.1 for case study data sets CS-MB.DAT, CS-VC.DAT, and CS-XS.DAT.

Exercise 8.3 Modify the grid in EXAM3.DAT so that it has only five blocks in the x direction, but the model volume is unchanged. Be sure to relocate the wells relative to the grid to keep them in their appropriate physical locations. How does the coarser grid affect model performance?

Exercise 8.4 Modify the grid in EXAM2.DAT so that it has $5 \times 5 \times 4$ grid blocks. The well should be in the center of the reservoir and the reservoir volume should be unchanged by the redefinition of the grid. How does the finer grid affect model performance when the model is run for three years?

References

Aziz, K. (1993): "Reservoir Simulation Grids: Opportunities and Problems," *Journal of Petroleum Technology*, pp. 658-663.

Aziz, K. and A. Settari (1979): **Petroleum Reservoir Simulation**, New York: Elsevier.

Chin, W.C. (1993): **Modern Reservoir Flow and Well Transient Analysis**, Houston: Gulf Publishing.

Christie, M.A. (Nov. 1996): "Upscaling for Reservoir Simulation," *Journal of Petroleum Technology*, pp. 1004-1010.

Dahlberg, E.C. (1975): "Relative Effectiveness of Geologists and Computers in Mapping Potential Hydrocarbon Exploration Targets," *Mathematical Geology*, Volume 7, pp. 373-394.

Englund, E. and A. Sparks (1991): "Geo-EAS 1.2.1 User's Guide," *Environmental Protection Agency Report #600/8-91/008 EPA-EMSL*, Las Vegas, NV.

Harpole, K.J. (1985): **Reservoir Environments and Their Characterization**, Boston: International Human Resources Development Corporation.

Harris, D.G. (May 1975): "The Role of Geology in Reservoir Simulation Studies," *Journal of Petroleum Technology*, pp. 625-632.

Hegre, T.M., V. Dalen, and A. Henriquez (1986): "Generalized Transmissibilities for Distorted Grids in Reservoir Simulation," Paper SPE 15622, *Proceedings of 61st Annual SPE Technical Conference and Exhibition*, Richardson, TX: Society of Petroleum Engineers.

Lee, S.H., L.J. Durlofsky, M.F. Lough, and W.H. Chen (1997): "Finite Difference Simulation of Geologically Complex Reservoirs with Tensor Permeabilities," Paper SPE 38002, *Proceedings of 1997 SPE Reservoir Simulation Symposium*, Richardson, TX: Society of Petroleum Engineers.

Mattax, C.C. and R.L. Dalton (1990): **Reservoir Simulation**, SPE Monograph #13, Richardson, TX: Society of Petroleum Engineers.

Pannatier, Y. (1996): **VARIOWIN: Software for Spatial Data Analysis in 2D**, New York: Springer-Verlag.

Raleigh, M. (Sept. 1991): *ECLIPSE Newsletter*, Houston: Schlumberger GeoQuest.

Taggart, I.J., E. Soedarmo, and L. Paterson (1995): "Limitations in the Use of Pseudofunctions for Up-Scaling Reservoir Simulation Models," Paper SPE 29126, *Proceeding of 13th Society of Petroleum Engineers Symposium on Reservoir Simulation*, San Antonio, TX, Feb. 12-15.

Tearpock, D.J. and R.E. Bischke (1991): **Applied Subsurface Geological Mapping**, Englewood Cliffs, NJ: Prentice Hall.

Todd, M.R., P.M. Odell, and G.J. Hiraski (Dec. 1972): "Methods for Increased Accuracy in Numerical Reservoir Simulators," *Society of Petroleum Engineering Journal*, pp. 515-530.

Young, L.C. (1984): "A Study of Spatial Approximations for Simulating Fluid Displacements in Petroleum Reservoirs," **Computer Methods in Applied Mechanics and Engineering**, New York: Elsevier, pp. 3-46.

Chapter 9

Data Preparation for a Typical Study

In a typical study it is necessary to first specify project objectives. The objectives help define the level of detail that will be incorporated in the reservoir model. Once objectives are defined, it is helpful to think of the study proceeding in three phases [Saleri, 1993]: the history match phase; a calibration phase, which provides a smooth transition between the first and third phases; and the prediction phase. The first step toward obtaining a history match is the collection and analysis of data.

9.1 Data Preparation

Data must be acquired and evaluated with a focus on its quality and the identification of relevant drive mechanisms that should be included in the model [for example, see Crichlow, 1977; Saleri, et al., 1992; Raza, 1992]. Given that information, it is possible to select the type of model that will be needed for the study: conceptual, window area, or full field model. In many cases all three of these models may need to be used, as illustrated in Fanchi, et al. [1996]. Data must be acquired for each model.

Some of the data that is required for a model study can be found in existing reports. The modeling team should find as many reports as it can from as many disciplines as possible. Table 9-1 lists the types of data that are needed in a model study. A review of geophysical, geological, petrophysical and engineering reports provides a background on how the project has been developed and what preconceived interpretations have been established. During the course of the study, it may be necessary to develop not only a new view of the reservoir, but also to prepare an explanation of why the new view is superior to a previously approved interpretation. If significant gaps exist in the reports, particularly historical performance of the field, it is wise to update them.

Table 9-1

Data Required for a Simulation Study

Property	Sources
Permeability	Pressure transient testing, Core analyses, Correlations, Well performance
Porosity, Rock compressibility	Core analyses, Well logs
Relative permeability and capillary pressure	Laboratory core flow tests
Saturations	Well logs, Core analyses, Pressure cores, Single-well tracer tests
Fluid property (PVT) data	Laboratory analyses of reservoir fluid samples
Faults, boundaries, fluid contacts	Seismic, Pressure transient testing
Aquifers	Seismic, Material balance calculations, Regional exploration studies
Fracture spacing, orientation, connectivity	Core analyses, Well logs, Seismic, Pressure transient tests, Interference testing, Wellbore performance
Rate and pressure data, completion and workover data	Field performance history

A review of rock and fluid property may show that the amount of available data is limited. If so, additional data should be obtained when possible. This may require special laboratory tests, depending on the objectives of the study. If measured data cannot be obtained during the scope of the study, then correlations or data from analogous fields will have to be used. Values must be entered into the simulator, and it is prudent to select values that can be justified.

Well data should be reviewed. If additional field tests are needed, they should be requested and incorporated into the study schedule. Due to the costs and operating constraints of a project, it may be difficult to justify the expense of acquiring more data or delaying the study while additional data is obtained. The modeling team should take care to avoid underestimating the amount of work that may be needed to prepare an input data set. It can take as long to collect and prepare the data as it does to do the study.

9.2 Pressure Correction

When pressures are matched in a model study, the calculated and observed pressures should be compared at a common datum. In addition, pressures from well tests should be corrected for comparison with model block pressures. A widely used pressure correction is the Peaceman [1978, 1983] correction.

Figure 9-1 illustrates a pressure buildup curve as a function of radial distance from the center of a wellbore with radius r_w. To obtain a well block pressure P_o from a pressure buildup (PBU), Peaceman used a Cartesian grid to model the PBU performance of a well to find an equivalent well block radius r_o. A Horner plot of a PBU test is illustrated in Figure 9-2.

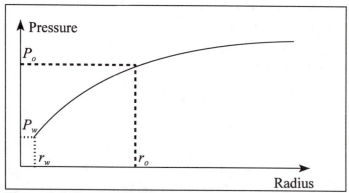

Figure 9-1. Pressure buildup.

Peaceman showed that the shut-in pressure P_{ws} of an actual well equals the simulator well block pressure P_o at a shut-in time Δt_s given by

$$\Delta t_s = \frac{1688\, \phi \mu c_T r_o^2}{K}$$

where K is permeability, ϕ is porosity, μ is viscosity, and c_T is total compressibility. Units for all variables are given in Table 9-2 at the end of this section.

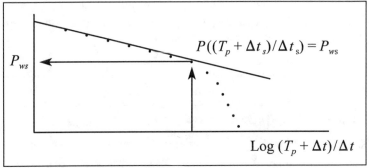

Figure 9-2 Horner Plot of PBU

The relationship between grid block pressure P_o and flowing pressure P_{wf} at the wellbore is

$$P_{wf} = P_o - 141.2\, \frac{QB\mu}{Kh} \left[\ln \frac{r_o}{r_w} + S \right]$$

where Q is the flow rate, B is formation volume factor, and S is skin. For an isotropic reservoir, that is, a reservoir in which x-direction permeability equals y direction permeability ($K_x = K_y$), the equivalent well block radius is given in terms of the block lengths $\{\Delta x, \Delta y\}$, thus

$$r_o = 0.14 \left(\Delta x^2 + \Delta y^2 \right)^{\frac{1}{2}}$$

shut-in time can be masked by wellbore storage effects. If it is, the shut-in pressure P_{ws} may have to be obtained by extrapolation of another part of the curve, such as the radial flow curve. Table 9-2 summarizes the parameters involved in the Peaceman correction for a consistent set of units. An application of the Peaceman correction is presented in Chapter 16.3 within the context of a case study. Peaceman's work with 2D models was extended to 3D by Odeh [1985].

Table 9-2

Oilfield Units for the Peaceman Correction

Parameter	Units
B	RB/STB
c_T	psi^{-1}
h	ft
K	md
P_o, P_{wf}, P_{ws}	psia
Q	STB/D
r_o, r_w	ft
S	fraction
Δt_s	hr
$\Delta x, \Delta y$	ft
ϕ	fraction
μ	cp

9.3 Simulator Selection and Ockham's Razor

Several requirements must be considered when selecting a simulator. These requirements can be classified into two general categories: reservoir and non-reservoir. From a reservoir perspective, we are interested in fluid type, reservoir architecture, and the types of recovery processes or drive mechanisms that are anticipated.

Reservoir architecture encompasses a variety of parameters that have a major impact on model design. Study objectives and the geologic model must be considered in establishing the dimensionality of the problem (1D, 2D, or 3D) and the geometry of the grid. Do we need special grid options, such as radial coning or local grid refinement, or will Cartesian coordinates be satisfactory? If the study is designed to investigate near wellbore flow, it would be wise to select a grid that provides good spatial resolution near the wellbore, for example, radial coordinates. On the other hand, if the study is intended to provide an overview of field performance, a coarse Cartesian grid may be satisfactory.

The level of complexity of the geology will influence grid definition and, in the case of fractured reservoirs, the type of flow equations that must be used [for example, see Reiss, 1980; Aguilera, 1980; Golf-Racht, 1982; and Lough, et al., 1996]. A highly faulted reservoir or a naturally fractured reservoir is more difficult to describe numerically than a homogeneous sand.

Model selection will be influenced by the types of processes and drive mechanisms that dominate flow in the reservoir. Processes range from gas cap drive and water drive under primary depletion, through water or gas injection in pressure maintenance programs, to miscible or thermal flooding in enhanced recovery projects. The choice of model will vary depending on the anticipated process. For example, dry gas injection in a condensate reservoir is typically modeled with a compositional simulator, while steam flooding a heavy oil reservoir should be modeled with a thermal simulator.

A few guidelines are worth noting with regard to simulator selection. Many novice modelers make the mistake of selecting models that are much more complex than they need to be to satisfy the objectives of the study. According to Coats [1969], the modeler should "select the least complicated model and grossest reservoir description that will allow the desired estimation of reservoir performance." This is a restatement of Ockham's Razor.

William of Ockham, a fourteenth-century English philosopher, said "plurality must not be posited without necessity" [Jefferys and Berger, 1992]. Today this is interpreted to mean that an explanation of the facts should be no more complicated than necessary. We should favor the simplest hypothesis that is consistent with the data.

Ockham's Razor should be applied with care, however, because one of the goals of a model study is to establish a consensus about how the reservoir behaves. This consensus is political, to an extent, because the model must satisfy the people who commissioned the study. Their views may require using a model that has more complexity than required from a technical modeling perspective.

Non-reservoir requirements include personnel, simulator availability, and cost effectiveness. Personnel will be needed to gather and evaluate data, prepare input data, perform the history match and then make predictions. Data gathering may take a few days or several months depending on the quality and extent of the data base for a particular field. The history matching and prediction phases do not necessarily have to be done by the same modeler. In some companies, history matching is done in a collaborative effort between a specialized technology center and a field office, while most of the prediction work is completed in the field office. This takes advantage of specialized expertise: technology centers, including outside consultants, routinely set up and run models, while day-to-day changes that impact production operations are handled in the field office. The division of labor between history matching and prediction makes sense in some circumstances.

A wide variety of simulators are available for a price. The work horse simulators – black oil and compositional – can often be leased on an as-needed basis or are available through computer networks. More specialized simulators may be obtained from software vendors, or as publicly available research codes developed at university and government laboratories.

As complexity increases, so also does cost. A good economic argument to support Ockham's Razor is to remember that the latest technology is not always the best technology for a project, and its use comes with a cost. Modeling teams are often tempted to apply the latest technology, even if it is not warranted. An example is the use of local grid refinement (LGR) to model horizontal wells. LGR is an innovative grid preparation technique that can improve spatial resolution, but at a substantial increase in computer cost and simulator sophistication. It is very common to find LGR used to model horizontal wells. In some cases, such as feasibility studies, this level of technical detail exceeds the needs of the study objectives and simply adds cost to the project without adding the corresponding value. A wise modeling team will match the level of technology with the objectives of the study. The result will be the selection of the most cost effective method for achieving study objectives.

The cost of a simulation study can be estimated based on previous experience with similar studies. As an example of how to estimate the cost for a black oil simulation study, begin by calculating GBTS, which is the product of the number of grid blocks and the number of time steps. Once GBTS is known, it should be related to the computer processing (cpu) time needed to make a run. The amount of cpu time per GBTS is determined by dividing the cpu time needed to make previous model runs by the number of GBTS in those runs. The product of GBTS and cpu time per GBTS gives total cpu time needed for a run. The cost of the study then depends on the number of runs that need to be made. The number of runs can be estimated by assuming that approximately 100 runs will be needed to obtain a history match. A similar approach is applied to estimating the cost of making predictions. Personnel cost is approximately equal to computer cost for the study. This does not include the cost of data collection and evaluation.

Exercises

Exercise 9.1 Data set EXAM10.DAT uses multiple Rock and PVT regions. Review EXAM10.DAT and simplify the data set without altering model results. Chapter 19.4 presents a description of Rock and PVT region data records.

Exercise 9.2 A model has 10 × 10 × 4 grid blocks and takes 5 minutes to run 100 time steps. Calculate cpu time per GBTS. Estimate how long it would take to make 100 runs with 200 time steps each.

References

Aguilera, R. (1980): **Naturally Fractured Reservoirs**, Tulsa, OK: PennWell Publishing.

Coats, K.H. (1969): "Use and Misuse of Reservoir Simulation Models," *Journal of Petroleum Technology*, pp. 183-190.

Crichlow, H.B. (1977): **Modern Reservoir Engineering – A Simulation Approach**, Englewood Cliffs, NJ: Prentice Hall.

Fanchi, J.R., H.-Z. Meng, R.P. Stoltz, and M.W. Owen (1996): "Nash Reservoir Management Study with Stochastic Images: A Case Study," *Society of Petroleum Engineers Formation Evaluation*, pp. 155-161.

Golf-Racht, T.D. van (1982): **Fundamentals of Fractured Reservoir Engineering**, New York: Elsevier.

Jefferys, W.H. and J.O. Berger (1992): "Ockham's Razor and Bayesian Analysis," *American Scientist*, Volume 80, pp. 64-72.

Lough, M.F., S.H. Lee, and J. Kamath (Nov. 1996): "Gridblock Effective-Permeability Calculation for Simulation of Naturally Fractured Reservoirs," *Journal of Petroleum Technology*, pp. 1033-1034.

Odeh, A.S. (Feb. 1985): "The Proper Interpretation of Field-Determined Buildup Pressure and Skin Values for Simulator Use," *Society of Petroleum Engineering Journal*, pp. 125-131.

Peaceman, D.W. (June 1978): "Interpretation of Well-Block Pressures in Numerical Reservoir Simulation," *Society of Petroleum Engineering Journal*, pp. 183-194.

Peaceman, D.W. (June 1983): "Interpretation of Well-Block Pressures in Numerical Reservoir Simulation with Nonsquare Grid Blocks and Anisotropic Permeability," *Society of Petroleum Engineering Journal*, pp. 531-543.

Raza, S.H. (1992): "Data Acquisition and Analysis for Efficient Reservoir Management," *Journal of Petroleum Technology*, pp. 466-468.

Reiss, L.H. (1980): **The Reservoir Engineering Aspects of Fractured Reservoirs**, Houston: Gulf Publishing.

Saleri, N.G. (1993): "Reservoir Performance Forecasting: Acceleration by Parallel Planning," *Journal of Petroleum Technology*, pp. 652-657.

Saleri, N.G., R.M. Toronyi, and D.E. Snyder (1992): "Data and Data Hierarchy," *Journal of Petroleum Technology*, pp. 1286-1293.

Chapter 10

History Matching

The history matching process begins with clearly defined objectives. Given the objectives, it is necessary to acquire model input data, especially the history of field performance. One of the essential tasks of the data acquisition stage is to determine which data should be matched during the history matching process. For example, if a gas-water reservoir is being modeled, gas rate is usually specified and water production is matched. By contrast, if an oil reservoir is being modeled, oil rate is specified and water and gas production are matched.

Data acquisition is an essential part of model initialization. Model initialization is the stage when the data is prepared in a form that can be used by the simulator. The model is considered initialized when it has all the data it needs to calculate fluids in place. The reservoir must be characterized in a format that can be put in a simulator and that is acceptable to the commissioners of the study. Reservoir characterization includes the selection of a grid and associated data for use in the model. It may also require the study of multiple reservoir realizations in the case of a geostatistical model study [for example, see Pannatier, 1996; Lieber, 1996; Rossini, et al., 1994; Englund and Sparks, 1991; Haldorsen and Damsleth, 1990; and Isaaks and Srivastava, 1989]. All fluid data corrections, such as flash corrections applied to differential PVT data in a black oil simulation, must be completed during the model initialization process.

In many cases, simple conceptual models may be useful in selecting a final grid for the model study, especially when determining the number of layers. As an illustration, suppose we want to track flood front movement in a very large field. In this case, we want as much areal definition as possible (at least 3 to 5 grid blocks between each grid block containing a well), but this may mean loss of vertical definition. A way to resolve the problem is to set up one or more cross-section models that represent different parts of the field. Vertical conformance effects in these regions are modeled in detail by calculating flow performance with the cross-section models. The flow performance of a detailed cross-section model is then matched by adjusting relative permeability curves in a model with fewer layers. The resulting pseudorelative permeability curves are considered acceptable for use in an areal model.

Another aspect of model initialization is equilibration. This is the point at which fluid contacts are established and fluid volumes are calculated. Resulting model volumes should be compared with other estimates of fluid in place, notably volumetric and material balance estimates. There should be reasonable agreement between the different methods (for example, within two percent). Finally, the history match can begin.

10.1 Illustrative History Matching Strategy

A universally accepted strategy for performing a history match does not exist. History matching is as much art as science because of the complexity of the problem. Nevertheless, there are some general guidelines that can help move a history match toward successful completion. These guidelines have been presented by such authors as Crichlow [1977], Mattax and Dalton [1990], Thomas [1982], and Saleri, et al. [1992]. One set of guidelines is presented in Table 10-1. The first two steps in the table take precedence over the last two. If the first two steps cannot be achieved, there is a good chance the model is inadequate and revisions will be necessary. An inadequate model may be due to a variety of problems: for example, the wrong model was selected, the reservoir is poorly characterized, or field data is inaccurate or incomplete.

Among the data variables matched in a typical black oil or gas study are pressure, production rate, WOR and GOR, and tracer data if it is available. More specialized studies, such as compositional or thermal studies, should also match data unique to the process, such as well stream composition or the temperature of produced fluids.

Table 10-1
Suggested History Matching Procedure

Step	Remarks
I	Match volumetrics with material balance and identify aquifer support.
II	Match reservoir pressure. Pressure may be matched both globally and locally. The match of average field pressure establishes the global quality of the model as an overall material balance. The pressure distribution obtained by plotting well test results at given points in time shows the spatial variation associated with local variability of field performance.
III	Match saturation dependent variables. These variables include water-oil ratio (WOR) and gas-oil ratio (GOR). WOR and GOR are often the most sensitive production variables, both in terms of breakthrough time and the shape of the WOR or GOR curve.
IV	Match well flowing pressures.

The pressure match is usually the first match to be sought during the history matching process. A comparison of estimated reservoir pressures obtained from well tests of a single well on successive days shows that errors in reported historical pressures can be up to 10 percent of pressure drawdown. This error may be as large or larger than the Peaceman correction discussed in Chapter 9. As a first approximation, it is sufficient to compare uncorrected historical pressures directly with model pressures, particularly if your initial interest is in pressure trends and not in actual pressure values. Pressure corrections should be applied when fine tuning the history match.

Production rates are usually from monthly production records. The modeler specifies one rate or well pressure, and then verifies that the rate is entered properly by comparing observed cumulative production with model cumulative production. After the rate of one phase is specified, the rates of all other phases must be matched by model performance. In many cases, observed rates will be averaged on a monthly or quarterly basis and then compared with model calculated rates. If the history of reservoir performance is extensive, then it is often wise to place a greater reliance on the validity of the most recent field data when performaing a history match.

Phase ratios, such as GOR and WOR, are sensitive indicators of model performance. Matching ratios provides information about pressure depletion and front movements. Tracers are also useful for modeling fluid fronts. Tracers need not be expensive chemicals; they can even be changes in the salinity of produced water. Salinity changes can occur as a result of mixing when injected and *in situ* brines have different salinities. Water sample analysis on a periodic basis is useful for tracking salinity variation as a function of time.

10.2 Key History Matching Parameters

A fundamental concept of history matching is the concept of a "hierarchy of uncertainty." The hierarchy of uncertainty is a ranking of model input data quality that lets the modeler determine which data is most and least reliable. Changes to model input data are then constrained by the principle that the least reliable data should be changed first. The question is: which data are least reliable?

Data reliability is determined when data is collected and evaluated for completeness and validity [Raza, 1992; Saleri, et al., 1992]. This is such an important step in establishing a feel for the data that the modeler should be closely involved with the review of data. Relative permeability data are typically placed at the top of the heirarchy of uncertainty because they are modified more often than other data. Relative permeability curves are often determined from core floods. As a consequence, the applicability of the final set of curves to the rest of the modeled region is always in doubt.

Initial fluid volumes may be modified by changing a variety of input parameters, including relative permeability endpoints and fluid contacts. Model calculated original fluid volumes in place are constrained by independent techniques like volumetrics and material balance studies.

Attempts to match well data may require changing the producing interval or the productivity index of a perforation interval. If it is difficult to match well performance in a zone or set of zones, the modeler needs to look at a variety of possibilities, including unexpected completion and wellbore problems. In one study, for example, an unexpectedly high GOR from a perforation interval that was known to be below the gas-oil contact was due to gas flow in the annulus between the tubing and the casing. This result was confirmed by running a cement bond log and finding a leak in the wellbore interval adjacent to the gas cap. Gas from the gas cap was entering the wellbore and causing the larger than expected production GOR. This effect can be modeled by a variety of options, depending on the degree of accuracy desired: for example, it could be modeled by altering productivity index (PI) in the well model or by designing a near wellbore conceptual model and preparing pseudorelative permeability curves. The choice of method will influence the predictive capability of the model. Thus, a pseudorelative permeability model will allow for high GOR even if the well is recompleted, whereas the PI could be readily corrected at the time of well recompletion to reflect the improvement in wellbore integrity.

Map adjustments may also be necessary. This used to be considered a last resort change because map changes required substantial effort to redigitize the modified maps and prepare a revised grid. Pre-processing packages and computer aided geologic modeling are making map changes a more acceptable history match method. In the case of geostatistics, a history matching process may actually involve the use of several different geologic models, called stochastic images. Geostatistics is discussed in more detail in Chapter 12.

Toronyi and Saleri [1988] present a detailed discussion of their approach to history matching. It is noteworthy because they provide guidance on how changes in some history match parameters affect matches of saturation and pressure gradients. A summary is presented in Table 10-2. It shows, for example, that a change in pore volume can effect pressure as it changes with time. As another example, relative permeability changes are useful for matching saturation variations in time and space. Notice that fluid property data are seldom changed to match field history. This is because fluid property data tend to be more accurately measured than other model input data.

History matching must not be achieved by making incorrect parameter modifications. For example, matching pressure may be achieved by adjusting rock compressibility, yet the final match value should be within the set of values typically associated with the type of rock in the formation. In general, modified parameter values must be physically meaningful.

Table 10-2
Influence of Key History Matching Parameters

Parameter	Pressure match	Saturation match
Pore volume	$\Delta P/\Delta t$	*
Permeability thickness	$\Delta P/\Delta x$	$\Delta S/\Delta x$
Relative permeability	Not used	$\Delta S/\Delta x$ and $\Delta S/\Delta t$
Rock compressibility	*	Not used
Bubble-point pressure	$\Delta P/\Delta t$ *	*
*Avoid changing if possible		

10.3 Evaluating the History Match

One way to evaluate the history match is to compare observed and calculated parameters. Typically, observed and calculated parameters are compared by making plots of pressure vs time, cumulative production (or injection) vs time, production (or injection) rates vs time, and GOR, WOR, or water cut vs time. Other comparisons can and should be made if data are available. They include, for example, model saturations versus well log saturations, and tracer concentration (such as salinity) versus time. In the case of compositional simulation, dominant components (typically methane) should be plotted as a function of time.

In many studies, the most sensitive indicators of model performance are plots of GOR, WOR, or water cut vs time. These plots can be used to identify problem areas. For example, suppose we plot all high/low WOR and GOR wells or plot all high/low pressure wells. A review of such plots may reveal a grouping of wells with the same problem. This can identify the presence of a systematic error or flaw in the model that needs to be corrected. If the distribution is random, then local variations in performance due to heterogeneity should be considered.

10.4 Deciding on a Match

There are several ways to decide if a match is satisfactory. In all cases, a clear understanding of the study objectives should be the standard for making the decision. If a coarse study is being performed, the quality of the match between observed and calculated parameters does not need to be as accurate as it would need to be for a more detailed study. For example, pressure may be considered matched if the difference between calculated and observed pressures is within ±10% drawdown. The tolerance of ±10% is determined by estimating the uncertainty associated with measured field pressures and the required quality of the study. A study demanding greater reliability in predictions may need to reduce the tolerance to ± 5% or even less, but it is unrealistic to seek a tolerance of less than one percent. The uncertainty applies not to individual well gauge pressures, which may be measured to a precision of less than one percent, but to estimates of average field or region pressure from two or more well tests. The latter error is generally much larger than the precision of a single well test. In any event, model calculated pressure trends should match field or region pressure performance.

Another sensitive indicator of the quality of a history match is the match of WOR, GOR or water cut. Three factors need to be considered: breakthrough time, the magnitude of the difference between observed and calculated values, and trends. Adjustments in the model should be made to improve the quality of each factor. Saleri [1993] has observed that a match of the field is more easily obtained than a match of individual well performance. Indeed, he notes that matching every well is virtually impossible. As a rule of thumb, the field match may be valid for a year or more without updating, and we can expect the well match to be valid for up to six months without updating. Deviations from this rule will vary widely, and will depend on the type of system modeled and the alignment of the interpreted model with reality. Indeed, gas reservoirs without aquifer influx may be accurately modeled for the life of the field, while a gas reservoir with complex lithology and water influx may never be satisfactorily matched.

Modelers must resist being drawn into the "one more run" syndrome. This occurs when a modeler (or member of the study team) wants to see "just one more run" to try an idea that has not yet been tried. In practice, a final match is often declared when the time or money allotted for the study is depleted.

10.5 History Match Limitations

History matching may be thought of as an inverse problem. An inverse problem exists when the dependent variable is the best known aspect of a system and the independent variable must be determined [Oreskes, et al., 1994]. For example, we know the production performance of the field, which is dependent on input variables such as permeability.

In the context of an inverse problem, the problem is solved by finding a set of reasonable reservoir parameters that minimizes the difference between model performance and historical performance of the field. As usual, we must remember that we are solving a non-unique problem whose solution is often as much art as science. The uniqueness problem arises from many factors. Most notable of these are unreliable or limited field data and numerical effects. Advances in hardware and software technology have made it possible to minimize the effects of numerical problems, or at least estimate their influence on the final history match solution. Data limitations are more difficult to resolve because the system is inherently underdetermined: we do not have enough data to be sure that our final solution is correct.

Test of Reasonableness

A model may be considered reasonable if it does not violate any known physical constraints. In many cases, a model may be acceptable if it is reasonable. In other situations, not only must physical constraints be satisfied, but approved processes for evaluating data must also be followed. Thus a model may be reasonable, but if it is based on an innovative technique that is reasonable but not approved, the model will be unacceptable. The modeler may use a method that is in the literature, but the commissioner of the study may have a philosophical or empirical objection to the method. Window area modeling is a good example of a method that may be reasonable but not acceptable because failure to adequately describe flux across window area boundaries can yield poor results. If someone in a position of authority or influence has had a bad experience with the modeling method, they may refuse to accept results from the model. Similarly, the modeler needs to be aware that some modeling methods are not universally accepted. At the very least, alternative methods may be needed to corroborate the disputed method as part of a sensitivity analysis or model validation exercise.

Exercises

Exercise 10.1 Multiply the pore volume of data set EXAM6.DAT by 0.9 and 1.1. How does the change in pore volume affect pressure as a function of time?

Exercise 10.2 Double the horizontal permeability in layer $K = 1$ of data set EXAM6.DAT. What is the effect on reservoir pressure and production, by layer, at the end of two years? File BTEMP.WEL provides rate information by layer for all wells.

Exercise 10.3 Set the x direction transmissibility to 0 between $I = 2$ and $I = 3$ for blocks ranging from $J = 1$ to $J = 4$ in layers $K = 1$ and $K = 2$ of data set EXAM6.DAT. This transmissibility barrier represents a flow barrier such as a sealing fault. How does the barrier alter flow patterns and the distribution of reservoir pressure?

References

Crichlow, H.B. (1977): **Modern Reservoir Engineering – A Simulation Approach**, Englewood Cliffs, NJ: Prentice Hall.

Englund, E. and A. Sparks (1991): "Geo-EAS 1.2.1 User's Guide," *Environmental Protection Agency Report #600/8-91/008 EPA-EMSL*, Las Vegas, NV.

Haldorsen, H.H. and E. Damsleth (April 1990): "Stochastic Modeling," *Journal of Petroleum Technology*, pp. 404-412.

Isaaks, E.H. and R.M. Srivastava (1989): **Applied Geostatistics**, New York: Oxford University Press.

Lieber, Bob (Mar/Apr 1996): "Geostatistics: The Next Step in Reservoir Modeling," *Petro Systems World*, pp. 28-29.

Mattax, C.C. and R.L. Dalton (1990): **Reservoir Simulation**, SPE Monograph #13, Richardson, TX: Society of Petroleum Engineers.

Oreskes, N., K. Shrader-Frechette, and K. Belitz (1994): "Verification, Validation, and Confirmation of Numerical Models in the Earth Sciences", *Science*, pp. 641-646, Feb. 4.

Pannatier, Y. (1996): **VARIOWIN: Software for Spatial Data Analysis in 2D**, New York: Springer-Verlag.

Raza, S.H. (1992): "Data Acquisition and Analysis for Efficient Reservoir Management," *Journal of Petroleum Technology*, pp. 466-468.

Rossini, C., F. Brega, L. Piro, M. Rovellini, and G. Spotti (Nov. 1994): "Combined Geostatistical and Dynamic Simulations for Developing a Reservoir Management Strategy: A Case History," *Journal of Petroleum Technology*, pp. 979-985.

Saleri, N.G. (1993): "Reservoir Performance Forecasting: Acceleration by Parallel Planning," *Journal of Petroleum Technology*, pp. 652-657.

Saleri, N.G., R.M. Toronyi, and D.E. Snyder (1992): "Data and Data Hierarchy," *Journal of Petroleum Technology*, pp. 1286-1293.

Thomas, G.W. (1982): **Principles of Hydrocarbon Reservoir Simulation**, Boston: International Human Resources Development Corporation.

Toronyi, R.M. and N.G. Saleri (1988): "Engineering Control in Reservoir Simulation," SPE 17937, *Proceedings of 1988 Society of Petroleum Engineers Fall Conference*, Oct. 2-5.

Chapter 11

Predictions

The previous chapters have shown how to build a working model of the reservoir and establish a level of confidence in the validity of model results. It is time to recall that modeling was undertaken to prepare a tool that would help us develop recommendations for a reservoir management program. The primary reservoir management objective is to determine the optimum operating conditions needed to maximize the economic recovery of hydrocarbons. This is accomplished, in principle, by marshaling accessible resources to

 ♦ optimize recovery from a reservoir, and
 ♦ minimize capital investments and operating expenses.

The commercial impact of the simulation study is the preparation of a cash flow prediction from projected field performance. Thus, the study is often completed by making field performance predictions.

11.1 Prediction Capabilities

Performance predictions are valuable for a variety of purposes. Predictions can be used to better interpret and understand reservoir behavior and they provide a means of determining model sensitivity to changes in input data. This sensitivity analysis can guide the acquisition of additional data for improving reservoir management.

Predictions enable people to estimate project life by predicting recovery vs time. Project life depends not only on the flow behavior of the reservoir, but also on commercial issues. Models let the user impose a variety of economic constraints on future reservoir performance during the process of estimating project life. These constraints reflect a range of economic criteria that will interest management, shareholders, and prospective investors.

Commercial interests are clearly important to the future of a project, and so are technical issues. It is often necessary to compare different recovery processes as part of a study. Since there is only one field, it is unrealistic to believe that many different recovery processes can be evaluated in the field, even as small-scale pilot projects. Pilot projects tend to be substantially more expensive to run than simulation studies. In some cases, however, it might be worthwhile to confirm a simulation study with a pilot project. This is especially true with expensive processes such as chemical and thermal flooding.

Yet another use for model predictions is the preparation of a reservoir management plan. Reservoir management plans have been discussed in previous sections. Their preparation is often the single most important motivation for performing a simulation study.

11.2 Prediction Process

The prediction process begins with model calibration. It is usually necessary to ensure continuity in well rate when the modeler switches from rate control during the history match to pressure control during the prediction stage of a study. This is illustrated in Figure 11-1 where the solid curve is the predicted rate based on the productivity index (PI) used in the history match. A clear discontinuity in rate is observed between the end of history and the beginning of prediction. The rate difference usually arises because the actual well PI, especially skin effect, is not accurately modeled by the model PI. An adjustment to model PI needs to be made to match final historical rate with initial predicted rate.

The next step is to prepare a base case prediction. The base case prediction is a forecast assuming existing operating conditions apply. For example, the base case for a newly developed field that is undergoing primary depletion should be a primary depletion case that extends to a user-specified economic limit. By contrast, if the field was being waterflooded, the waterflood should be the base case and alternative strategies may include gas injection and WAG (water-alternating-gas).

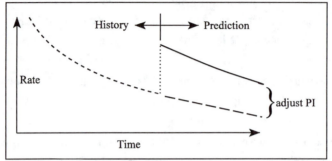

Figure 11-1. Model calibration.

The base case prediction establishes a basis from which to compare changes in field performance resulting from changes in existing operating conditions. In addition, a sensitivity analysis should be performed to provide insight into the uncertainty associated with model predictions. A procedure for conducting a sensitivity analysis is outlined below.

---

11. 3 Sensitivity Analyses

Sensitivity analyses are often needed in both the history matching and prediction stages [for example, see Crichlow, 1977; Mattax and Dalton, 1990; Saleri, 1993; and Fanchi, et al., 1996]. Any method that quantifies the uncertainty or risk associated with selecting a particular prediction case may be viewed as a sensitivity analysis. An example of a sensitivity analysis technique that is cost-effective in moving a history match forward is conceptual modeling. It can be used to address very specific questions, such as determining the impact of fluid contact movement on hydrocarbon recovery. Similarly, window models that study such issues as the behavior of a horizontal well in a fault block provide useful information on the sensitivity of a model to changes in input parameters.

Another example of a sensitivity analysis technique is risk analysis. Murtha [1997] defines risk analysis as "any form of analysis that studies and hence attempts to quantify risks associated with an investment." Risk in this context refers to a potential "change in assets associated with some chance occurrences." Risk analysis generates probabilities associated with changes of model input parameters. The parameter changes must be contained within ranges that are typically determined by the range of available data, information from analogous fields, and the experience of the modeling team. Each model run using a complete set of model input parameters constitutes a trial. A large number of trials can be used to generate probability distributions. Alternatively, the results of the trials can be used in a multivariable regression analysis to generate analytical expressions, as described below.

One of the most widely used techniques for studying model sensitivity to input parameter changes is to modify model input parameters in the history matched model. The following procedure combines multivariable regression and the results of model trials to generate an analytical expression for quantifying the effect of changing model parameters.

Assume a dependent variable F has the form

$$F = \kappa \prod_{j=1}^{n} X_j^{e_j}$$

where $\{X_j\}$ are n independent variables and κ is a proportionality constant that depends on the units of the independent variables. Examples of X_j are well separation, saturation end points, and aquifer strength. Taking the logarithm of the defining equation for F linearizes the function F and makes it suitable for multivariable regression analysis, thus

$$\ln F = \ln \kappa + \sum_{j=1}^{n} e_j \ln X_j$$

A sensitivity model is constructed using the following procedure:

♦ Run a model with different values of $\{X_j\}$
♦ Obtain values of F for each set of values of $\{X_j\}$

The constants κ, $\{e_j\}$ are obtained by performing a multivariable regression analysis using values of F calculated from the model runs as a function of $\{X_j\}$.

In addition to quantifying behavior, the regression procedure provides an estimate of fractional change of the dependent variable F when we make fractional changes to the independent variables $\{X_j\}$. The fractional change in F is given by

$$\frac{dF}{F} = \sum_{j=1}^{n} e_j \frac{dX_j}{X_j}$$

This lets us compare the relative importance of changes to the independent variables. Notice that the proportionality constant κ has been factored out of the expression dF/F for the fractional change in F. Thus, the quantity dF/F does not depend on the system of units used in the sensitivity study.

11.4 Economic Analysis

In addition to providing technical insight into fluid flow performance, model predictions are frequently combined with price forecasts to estimate how much revenue will be generated by a proposed reservoir management plan. The revenue stream is used to pay for capital and operating expenses, and the economic performance of the project depends on the relationship between revenue and expenses [see, for example, Bradley and Wood, 1994; Mian, 1992].

In a very real sense, the reservoir model determines how much money will be available to pay for wells, compressors, pipelines, platforms, processing facilities, and any other items that are needed to implement the plan represented by the model. For this reason, the modeling team may be expected to generate flow predictions using a combination of reservoir parameters that yield better recoveries than would be expected if a less "optimistic" set of parameters had been used. The sensitivity analysis is a useful process for determining the likelihood that a set of parameters will be realized. Indeed, modern reserves classification systems are designed to present reserves estimates in terms of their probability of occurrence.

According to the Society of Petroleum Engineers and the World Petroleum Congress [Staff-JPT, 1997], reserves are those quantities of petroleum which are anticipated to be commercially recoverable from known accumulations from a given date forward. Table 11-1 summarizes the SPE/WPC definitions of reserves. A different, albeit analogous, set of definitions exists in the Russian Federation [Nemchenko, et al., 1995; Grace, et al., 1993].

Table 11-1

SPE/WPC Reserves Definitions

Proved reserves	♦ Those quantities of petroleum which, by analysis of geological and engineering data, can be estimated with reasonable certainty to be commercially recoverable, from a given date forward, from known reservoirs and under current economic conditions, operating methods, and government regulation. ♦ In general, reserves are considered proved if the commercial producibility of the reservoir is supported by actual production or formation tests. ♦ There should be at least a 90% probability (P_{90}) that the quantities actually recovered will equal or exceed the estimate.
Unproved reserves	Those quantities of petroleum which are based on geologic and/or engineering data similar to that used in estimates of proved reserves; but technical, contractual, economic, or regulatory uncertainties preclude such reserves being classified as proved.
Probable reserves	♦ Those unproved reserves which analysis of geological and engineering data suggests are more likely than not to be recoverable. ♦ There should be at least a 50% probability (P_{50}) that the quantities actually recovered will equal or exceed the estimate.
Possible reserves	♦ Those unproved reserves which analysis of geological and engineering data suggests are less likely to be recoverable than probable reserves. ♦ There should be at least a 10% probability (P_{10}) that the quantities actually recovered will equal or exceed the estimate.

The probability distribution associated with the SPE/WPC reserves definitions can be estimated with relative ease if the modeling team has performed a sensitivity analysis that generates a set of cases that yield low, medium, and high reserves estimates. In the absence of data to the contrary, a reasonable first approximation is that each case is equally likely to occur. Given this assumption, an average μ and standard derivation σ may be calculated from the sensitivity analysis results to prepare a normal distribution of reserves. The resulting distribution can be used to associate an estimate of the likelihood of occurrence of any particular prediction case with its corresponding economic forecast. For a normal distribution with mean μ and standard deviation σ, the SPE/WPC reserves definitions are quantified as follows:

Proved reserves $= P_{90} = \mu - 1.28\sigma$
Probable reserves $= P_{50} = \mu$
Possible reserves $= P_{10} = \mu + 1.28\sigma$

A probabilistic analysis gives decision making bodies such as corporate managements and financial institutions the information they need to make informed decisions.

11.5 Validity of Model Predictions

The validity of model predictions was studied by Saleri [1993] who compared actual field performance with predicted performance. Figure 11-2 illustrates his results. The overall match of field performance, such as total rate and pressure performance, is reasonable. The field match is somewhat deceptive however, because the validity of individual well performance forecasting varies widely. Indeed, the match of water and gas performance for about half of the wells was deemed a "bust" by the author. This is not unusual in a model study. Saleri arrived at the following conclusions:

♦ "Barring major geologic and/or reservoir data limitations, fieldwide cumulative production forecast accuracies would tend to range from 10% to 40%." [Saleri, 1993]

♦ "Well performance forecasts are bound to be less successful than fieldwide predictions." [Saleri, 1993]

These points underscore the need to recognize that the history match process does not yield a unique solution. Forecasts of reservoir behavior depend on the validity of the history match.

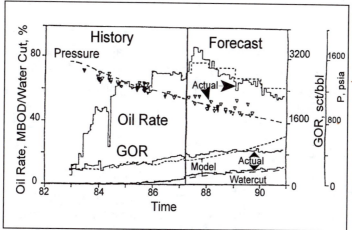

Figure 11-2. Quality of field performance match (after Saleri, 1993; reprinted by permission of the Society of Petroleum Engineers).

Despite the uncertainty associated with simulator-based forecasts, reservoir simulation continues to be the most reliable method for making performance predictions, particularly for reservoirs that do not have an extensive history or for fields that are being considered as candidates for a change in reservoir management strategy. Other methods, such as decline curve analysis and material balance analysis, can generate performance forecasts, but not to the degree of detail provided by a reservoir model study. As Saleri [1993] noted,

♦ "While a 10% to 40% forecast uncertainty may appear alarming in an absolute sense, the majority of reservoir engineering decisions require choices based solely on comparative analyses (for example, peripheral vs. pattern flood). Thus, in selecting optimum management strategies, finite-difference models still offer the most effective tools."

Saleri's view is similar to that of Oreskes, et al. [1994]. Even though models are non-unique representations of nature, they still have many uses. In summary, models can be used to

♦ corroborate or refute hypotheses about physical systems;

♦ identify discrepancies in other models; and

♦ perform sensitivity analyses.

Part II integrates the ideas presented above in the context of a case study.

Exercises

Exercise 11.1 Data set EXAM4.DAT is a 2D areal model of an undersaturated oil reservoir undergoing primary depletion. Set the bottom hole pressure (BHP) in well P-1 of EXAM4.DAT to 150 psia and run the model. How much oil is recovered?

Exercise 11.2 Beginning at the end of year one, add a water injection well in each of the four corner grid blocks in data set EXAM4.DAT with the BHP modification described in Exercise 11.1. The maximum allowable BHP for an injection well is 5000 psia. Assume the target rate for the oil production well is 600 STB/D. Maximize oil recovery by varying the amount of water injected. Data set EXAM6.DAT is an example of a data set with the injection wells added.

References

Bradley, M.E. and A.R.O. Wood (Nov. 1994): "Forecasting Oilfield Economic Performances," *Journal of Petroleum Technology*, pp. 965-971 and references therein.

Crichlow, H.B. (1977): **Modern Reservoir Engineering – A Simulation Approach**, Englewood Cliffs, NJ: Prentice Hall.

Fanchi, J.R., H.-Z. Meng, R.P. Stoltz, and M.W. Owen (1996): "Nash Reservoir Management Study with Stochastic Images: A Case Study," *Society of Petroleum Engineers Formation Evaluation*, pp. 155-161.

Grace, J.D., R.H. Caldwell, and D.I. Heather (Sept. 1993): "Comparative Reserves Definitions: USA, Europe, and the Former Soviet Union," *Journal of Petroleum Technology*, pp. 866-872.

Mattax, C.C. and R.L. Dalton (1990): **Reservoir Simulation**, SPE Monograph #13, Richardson, TX: Society of Petroleum Engineers.

Mian, M.A. (1992): **Petroleum Engineering Handbook for the Practicing Engineer**, Volume 1, Tulsa, OK: PennWell Publishing.

Murtha, J.A. (1997): "Monte Carlo Simulation: Its Status and Future," *Journal of Petroleum Technology*, pp. 361-373.

Nemchenko, N.N., M.Y. Zykin, A.A. Arbatov, V.I. Poroskun, and I.S. Gutman (1995): "Distinctions in the Oil and Gas Reserves and Resources Classifications Assumed in Russia and USA – Source of Distinctions," *Energy Exploration and Exploitation*, Volume 13, #6, Essex, United Kingdom: Multi-Science Publishing Company.

Oreskes, N., K. Shrader-Frechette and K. Belitz (1994): "Verification, Validation, and Confirmation of Numerical Models in the Earth Sciences", *Science*, pp. 641-646, Feb. 4.

Saleri, N.G. (1993): "Reservoir Performance Forecasting: Acceleration by Parallel Planning," *Journal of Petroleum Technology*, pp. 652-657.

Staff-JPT (May 1997): "SPE/WPC Reserves Definitions Approved," *Journal of Petroleum Technology*, pp. 527-528.

Chapter 12

The Future of Reservoir Modeling

The process of characterizing a reservoir in a format that is suitable for use in a reservoir simulator begins with the gathering of data at control points such as wells (Figure 8-1). Once this occurs, the data can be contoured and digitized. The resulting set of digitized maps becomes part of the input data set for a reservoir simulator.

The contouring step in the process outlined above is changing. Contouring is the step in which reservoir parameters such as thickness and porosity are spatially distributed. The spatial distribution of reservoir parameters is a fundamental aspect of the reservoir characterization process. Two methods for spatially distributing reservoir parameters are emerging: geostatistics and reservoir geophysics.

Many modelers view geostatistics as the method of choice for sophisticated reservoir flow modeling [for example, see Lieber, 1996; Haldorsen and Damsleth, 1993; and Rossini, et al., 1994], even though the resulting reservoir characterization is statistical. By contrast, information obtained from reservoir geophysics is improving our ability to "see" between wells in a deterministic sense. Are these methods competing or complementary? The answer to this question will exert a major influence on the future of reservoir modeling.

This chapter contains two case studies that demonstrate several points about geostatistics and reservoir geophysics. A review of these studies can help you decide whether either method is appropriate for a particular application. The case studies help put geostatistics and reservoir geophysics in perspective relative to state-of-the-art reservoir modeling.

12.1 Geostatistical Case Study

An example of a full field model study using a geostatistical reservoir realization is the reservoir management study of the N.E. Nash Unit in Oklahoma [Fanchi, et al., 1996]. The goal of the study was to prepare a full field reservoir model that could be used to identify unswept parts of the field. We knew, based on the history of the field, that water was breaking through at several wells. The study was designed to look for places where an additional production well could be economically drilled.

The N.E. Nash Unit has a gradual dip from north to south. The Misener sandstone reservoir is bounded above by the Woodford shale, on the flanks by the Sylvan shale, and below by the Viola limestone. The Viola limestone does allow some aquifer support for the Misener sandstone.

One of the primary tasks of the study was to map the N.E. Nash Unit. Two sets of maps were prepared: conventional hand-drawn maps, and a set of maps based on a geostatistical analysis of the field. The hand-drawn maps correspond to the deterministic approach in which a single realization is used, while the geostatistical maps correspond to a stochastic image of the reservoir.

A geostatistical analysis was performed using 42 well control points to calculate structural tops, gross thickness, net-to-gross ratio, and porosity. A cross-plot between porosity and core permeability yielded a relationship for calculating permeability from porosity. From this data, directional semi-variograms (Table 12-1) were prepared to describe the spatial continuity of each parameter. The semi-variograms represent parameter changes as functions of distance and direction. For a detailed technical discussion of geostatistics, see a text such as Isaaks and Srivastava [1989]. Hebert, et al. [1993] have published some geostatistical software that is compatible with BOAST II.

When two sets of maps were compared, the hand-drawn maps were found to be more homogenous than the geostatistical maps. The geostatistical maps exhibited the large scale trends shown in the hand-drawn maps, but contained more local variability. This was not surprising, since additional heterogeneity is expected to arise as a result of geostatistical mapping.

Table 12-1
Semi-Variogram Model

Goal: Model spatial correlation of data with semi-variance $\gamma(h)$	
Semi-Variance $$\gamma(h) = \frac{1}{2N(h)} \sum_{i=1}^{N(h)} \left[Z(x_i) - Z(x_i + h) \right]^2$$	
$Z(x_i)$	Value of spatially distributed property at point x_i, for example, ϕ, K.
h	Spatial vector or "lag" distance between data point at $x_i + h$ and data point at x_i. "Lag" h is a vector with length and direction.
$N(h)$	Number of data pairs approximately separated by vector h.

The choice of final maps was based on management priorities: minimize the risk of drilling a dry hole on the flanks of the field, and complete the study before water breakthrough occurred in the remaining oil producers. The geostatistical model satisfied both of these criteria. The main flow path in the reservoir was narrower in the geostatistically generated maps than in the hand-drawn maps, and the geostatistical realization could be modified in a day or two.

Once a set of maps was chosen, the history match process could begin. Tracer information in the form of salinity changes was useful in helping identify sources of injection water as the water was produced. This was valuable in defining flow channels that could not otherwise be inferred. In some areas, transmissibility and porosity changes were needed to match water cut and reservoir pressure.

The geostatistical realization used in the N.E. Nash study was just a single realization. It was selected because it satisfied constraints imposed by previous volumetric and material balance studies. If these constraints were not available or were less reliable, which would be the case early in the life of a field, a geostatistical study would require the use of multiple realizations to characterize the reservoir. This raises the question of how many realizations are necessary.

Figure 12-1 shows a random sampling from a discrete probability distribution. A running average is also plotted. The figure shows that the running average does not stabilize, or approach a constant value, until at least 20 trials have been completed. This is a large number of realizations if history matching is needed for each realization. Indeed, it would be an unacceptably large number of realizations, in most cases, because of the time it takes to perform a history match. There is no established procedure for selecting one or more realizations for history matching from a set of geostatistically derived realizations. One procedure is described by Rossini, et al. [1994].

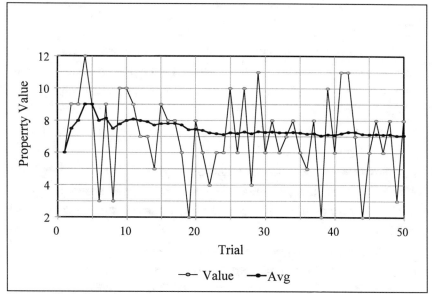

Figure 12-1. Running average.

Multiple realizations can also confuse people who are not closely in-volved with the modeling process because they do not have a single picture of the reservoir. On the other hand, the use of multiple realizations makes it possible to quantify the uncertainty associated with our limited knowledge of properties distributed spatially throughout the field.

Table 12-2 summarizes the advantages and concerns associated with geostatistics.

Table 12-2

Geostatistics

Advantages	Concerns
♦ Realism ♦ Quantifies uncertainty	♦ Cost and confusion of multiple realizations ♦ History matching still necessary to account for model discontinuities such as channeling ♦ History matching more complex due to factors such as probabilistically generated heterogeneity

12.2 Reservoir Description Using Seismic Data

The use of reservoir geophysics in the reservoir characterization process was introduced in Chapter 4. Reservoir geophysics has the potential to image important reservoir parameters in regions between wells. This potential has limitations, but before discussing these limitations, let us first consider how reservoir geophysics may be used and review an example where the potential of reservoir geophysics was realized.

The reservoir geophysical procedure requires the correlation of seismic data with reservoir properties. Correlations are sought by making crossplots of seismic data with reservoir properties. Some correlation pairs are listed below.

♦ Seismic Amplitude vs Rock Quality
  ◊ Rock Quality = kh_{net}, ϕkh_{net}, etc.
♦ Seismic Amplitude vs Oil Productive Capacity (OPC)
  ◊ OPC = $S_o \phi kh_{net}$
♦ Acoustic Impedance vs Porosity

If a statistically significant correlation is found, it can be used to guide the distribution of reservoir properties between wells. Ideally, the property distribution procedure will preserve reservoir properties at wells.

De Buyl et al. [1988] used reservoir geophysics to predict reservoir properties of two wells. They correlated well log derived properties with seismically controlled properties, for example, porosity, then used the correlation to distribute properties. Maps drawn from seismically controlled distributions exhibited more heterogeneity than conventional maps drawn from well log derived properties. Unlike geostatistics, where additional heterogeneity is obtained by sampling from a probability distribution, heterogeneity based on seismically controlled distributions represents spatial variations in reservoir properties determined by direct observation, albeit observation based on interpreted seismic data.

An indication of the technical success of the reservoir geophysical technique is given in Table 12-3. Actual values of reservoir parameters at two well locations are compared with values predicted using both well log derived properties and seismically controlled properties. This work by De Buyl, et al. [1988] is notable because it scientifically tests the seismic method: it makes predictions and then uses measurements to assess their validity. In this particular case, a reservoir characterization based on seismically controlled properties yielded more accurate predictions of reservoir properties than predictions made using a reservoir characterization based only on well data.

Table 12-3
Predictions at New Wells from Seismic and Well Data
[de Buyl, et al., 1988]

Well		Measured Values	Seismic Predicted	Well Data Predicted
I	Top of Reservoir (m)	-178.0	-175.0	-181.0
	Gross Porosity (vol %)	15.0	15.5	15.4
	Net ϕh (m)	1.78	1.53	1.96
J	Top of Reservoir (m)	-182.0	-179.0	-174.0
	Gross Porosity (vol %)	13.9	10.6	8.0
	Net ϕh (m)	1.08	1.05	0.15

Although reservoir geophysical techniques are still evolving, it is possible to make some general statements about the relative value of this emerging technology. Table 12-4 summarizes the advantages and concerns associated with reservoir geophysics.

Table 12-4
Reservoir Geophysics

Advantages	Concerns
◆ Able to "see" between wells ◆ Single realizations enhance 　◇ communication 　◇ understanding	◆ Cost of data acquisition and analysis ◆ Limited applicability ◆ Validity of realization unknown without sensitivity analysis

To demonstrate the limits of applicability of reservoir geophysics, the reservoir geophysical algorithm in BOAST4D (Chapter 30) was used to study a hypothetical reservoir system in which we could expect to see significant changes in seismic properties as a function of field performance over time. In particular, a dipping gas reservoir with aquifer influx was studied. The reservoir grid is shown in Figure 12-2. The reservoir has an initial gas saturation of 70% and an initial irreducible water saturation of 30%. The initial ratio of compressional velocity to shear velocity (V_p/V_s) was 1.684. A downdip

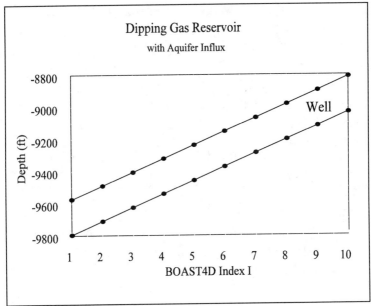

Figure 12-2. Cross-section of dipping gas reservoir.

aquifer provides pressure support and water invasion as the reservoir is produced.

Figure 12-3 shows the results after one year of depletion with aquifer influx for a system with an irreducible gas saturation (Sgr) of 0%. The change in gas saturation shows the influx of aquifer water. The change in fluid content changes fluid bulk modulus. As a consequence, the ratio V_p/V_s changes significantly in the waterflooded part of the reservoir.

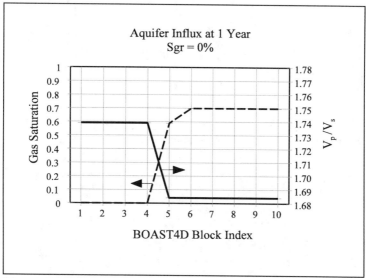

Figure 12-3. Reservoir performance with Sgr = 0%.

If we rerun the example with an irreducible gas saturation of three percent, we obtain the results shown in Figure 12-4. The large change in V_p/V_s is no longer observed because the presence of a small amount of gas significantly changed the compressibility of the system.

Time-lapse seismic tomography, or 4D seismic, could be used in our hypothetical example to track the movement of invading aquifer water, but the presence of a small amount of gas in the invaded zone increases the difficulty of detecting the gas-water contact. Calculations of 4D seismic performance based on algorithms like the one coded in BOAST4D can predict 4D seismic responses, but such algorithms are not yet widely available in commercial simulators.

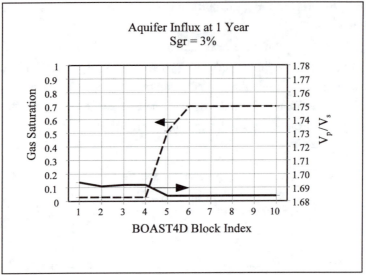

Figure 12-4. Reservoir performance with Sgr = 3%.

12.3 Trends in Simulation Technology

Although it is risky to predict technological developments, it is possible to infer trends by extrapolating ongoing research activities in the industry. A few industry leaders have made such extrapolations. Thakur [1996], for example, writes that data management and the integration of disciplines will play an increasingly important role in the future of reservoir modeling. Many modelers believe that the integration of disciplines will occur in reservoir modeling as the incorporation of more seismic and geological detail in models with finer 3D grids [He, et al., 1996; Kazemi, 1996; Uland, et al., 1997]. Indeed, He, et al. of Lamont-Doherty Earth Observatory stated that "4D, or time-dependent, seismic reservoir monitoring is an emerging technology that holds great hope as an oil-production management system."

These views should be kept in perspective. The choice of a reservoir characterization method depends on the applicability of the method to a given formation and its associated fluid properties. The cost of applying highly sophisticated simulation technology to a field should also be justified by the expected benefits. Simulators with appropriate algorithms can be used to assess the benefits and applicability of a characterization method to a particular situation.

Exercises

Exercise 12.1 Roll a pair of dice 50 times and record the results. Calculate a running average by calculating a new average after each trial (roll of the dice). Plot the running average for each trial. How many trials are necessary before the average stabilizes, that is, the average approaches a constant value?

Exercise 12.2 Review Exercises 4.1 and 4.2 in Chapter 4. Plot gas saturation and the ratio V_p/V_s versus BOAST4D index I. The results should be equivalent to the plots in Figures 12-3 and 12-4.

References

de Buyl, M., T. Guidish, and F. Bell (1988): "Reservoir Description from Seismic Lithologic Parameter Estimation", *Journal of Petroleum Technology*, pp. 475-482

Fanchi, J.R., H.Z. Meng, R.P. Stoltz, and M.W. Owens (1996): "Nash Reservoir Management Study with Stochastic Images: A Case Study," *Society of Petroleum Engineers Formation Evaluation*, pp. 155-161.

Haldorsen, H.H. and E. Damselth (1993): "Challenges in Reservoir Characterization," *American Association of Petroleum Geologists Bulletin* Volume 77, No. 4, pp. 541-551.

He, W., R.N. Anderson, L. Xu, A. Boulanger, B. Meadow, and R. Neal (1996): "4D Seismic Monitoring Grows as Production Tool," *Oil & Gas Journal*, pp. 41-46, May 20.

Hebert, H., A.T. Bourgoyne, Jr., and J. Tyler (May 1993): "BOAST II for the IBM 3090 and RISC 6000", U.S. Department of Energy Report DOE/ID/12842-2, Bartlesville Energy Technology Center, OK.

Isaaks, E.H. and R.M. Srivastava (1989): **Applied Geostatistics**, New York: Oxford University Press.

Kazemi, H. (Oct. 1996): "Future of Reservoir Simulation", *Society of Petroleum Engineers Computer Applications*, pp. 120-121.

Lieber, Bob (Mar/Apr 1996): "Geostatistics: The Next Step in Reservoir Modeling," *Petro Systems World*, pp. 28-29.

Rossini, C., F. Brega, L. Piro, M. Rovellini, and G. Spotti (Nov. 1994): "Combined Geostatistical and Dynamic Simulations for Developing a Reservoir Management Strategy: A Case History," *Journal of Petroleum Technology*, pp. 979-985.

Thakur, G.C. (1996): "What Is Reservoir Management?," *Journal of Petroleum Technology*, pp. 520-525.

Uland, M.J., S.W. Tinker, and D.H. Caldwell (1997): "3-D Reservoir Characterization for Improved Reservoir Management," paper presented at the 1997 Society of Petroleum Engineers' 10th Middle East Oil Show and Conference, Bahrain (March 15-18).

Part II

Case Study

Chapter 13

Study Objectives and Data Gathering

13.1 Study Objectives

The first step in a study is to identify its objectives. Two objectives of the case study are to increase understanding of the reservoir simulation process and to acquire experience working with a simulator. The experience you gained working the exercises in Part I is a transferrable skill. Many of the tasks performed with BOAST4D may differ in detail from other simulators, but are conceptually universal. Although the above objectives are important from a pedagogical point of view, they are secondary within the context of the case study.

The reservoir management objective of the case study is to *optimize production from a dipping, undersaturated oil reservoir.* There will be constraints imposed on the case study objective. Before discussing the constraints, however, it is first necessary to gather some background information about the field.

13.2 Reservoir Structure

A seismic line through an east-west cross-section of the field is shown in Figure 13-1. The broken lines in the figure are seismic reflectors and represent changes in acoustic impedance. The single well (P-1) has been producing from what appears to be a fault block bounded upstructure and to the east by an unconformity; downstructure and to the west by a fault or aquifer; and to the north and south by sealing faults.

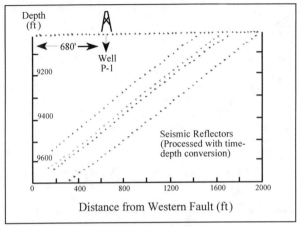

Figure 13-1. East-west seismic line.

A well log trace is shown in Figure 13-2. The well log trace provides a control for determining the time-depth conversion used in Figure 13-1. An analysis of the well log data shows that two major sands are present and are separated by a shale section. This is consistent with the seismic reflectors shown in Figure 13-1. Well log results are presented in Table 13-1. The table headings refer to porosity ϕ, water saturation S_w, gross thickness h, and the net-to-gross ratio NTG.

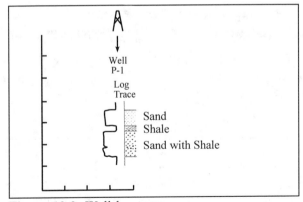

Figure 13-2. Well log trace.

Table 13-1
Well Log Analysis Summary

Lithology (From Cuttings)	Depth (ft) to Top of Formation	ϕ (fr.)	S_w (fr.)	h (ft)	NTG (fr.)
Sandstone	9330	0.20	0.30	80	0.9
Shale	9410	—	—	20	—
Sandstone with Shale Stringer	9430	0.25	0.30	120	0.8

A conceptual sketch of the reservoir cross-section is shown in Figure 13-3. Notice that we have adopted an unconformity as our geologic model.

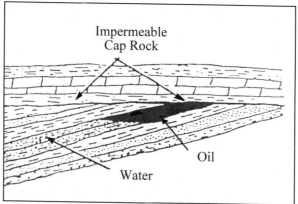

Figure 13-3. Conceptual sketch of reservoir cross-section (after Clark, 1969; reprinted by permission of the Society of Petroleum Engineers).

13.3 Production History

Well P-1 has produced for a year. Its production history is shown in Tables 13-2a and 13-2b.

Table 13-2a
Production History

TIME	OIL	GAS	WATER	GOR	WOR
		RATES			
DAYS	STB/D	MSCF/D	STB/D	SCF/STB	
5	500	228	0	457	0
13	500	228	0	457	0
24	500	228	0	457	0
41	500	228	0	457	0
66	500	228	0	457	0
91	500	228	1	457	0
122	500	228	1	457	0
153	500	228	1	457	0
183	500	228	2	457	0
214	500	228	2	457	0
245	500	228	2	457	0
274	500	228	3	457	0
305	500	228	3	457	0
336	500	228	4	457	0
365	500	228	5	457	0

A review of Tables 13-2a and 13-2b shows that oil rate has remained constant. Real data would show some variability, of course, but we are using an idealized data set to simplify the problem. Gas production has also remained constant, and there has been no change in the gas-oil ratio. This suggests that the reservoir is undersaturated, that is, reservoir pressure is above bubble point pressure so that only one hydrocarbon phase – the liquid phase – existed in the reservoir initially.

Table 13-2b

Production History

TIME	AVG RES PRESSURE	CUM PROD		
		OIL	GAS	WATER
DAYS	PSIA	MSTB	MMSCF	MSTB
5	3929	3	1	0
13	3915	6	3	0
24	3906	12	5	0
41	3901	20	9	0
66	3899	33	15	0
91	3898	46	21	0
122	3897	61	28	0
153	3897	77	35	0
183	3897	91	42	0
214	3896	107	49	0
245	3895	122	56	0
274	3895	137	63	0
305	3894	152	70	0
336	3893	168	77	1
365	3892	183	83	1

13.4 Drill Stem Test

Well P-1 logs and cores showed the presence of two major sands. A drill stem test (DST) was subsequently run in each major sand. Basic facts from the DST are summarized in Table 13-3.

Table 13-3

Summary of Well P-1 DST Results

Wellbore Radius	0.25 ft
Wellbore Skin	-0.5
Initial Pressure	3936 psia at 9360 ft
No-Flow Boundary	Within 700 ft

Permeability was estimated from the DST data for each sand. The results, together with average water saturation (S_w) values and calculated oil saturation (S_o) values, are presented in Table 13-4 for both major sands.

Table 13-4

Saturation and Permeability Values for Each Major Sand

Sand	S_w	$S_o = 1 - S_w$	Permeability (md)
1	0.3	0.7	75
2	0.3	0.7	250

13.4.1 DST Radius of Investigation

The radius of investigation for a DST can be estimated at various shut-in times by using the formula

$$r_i = 0.029 \sqrt{\frac{K \Delta t}{\phi \mu c_T}}$$

where K is permeability in md, ϕ is fractional porosity, μ is viscosity in cp, c_T is total compressibility in 1/psia, and Δt is shut-in time in hours. The following physical properties apply to the case study DST:

K	permeability	250 md
ϕ	porosity	0.228
μ	viscosity	0.71 cp
c_T	total compressibility	13×10^{-6} psia^{-1}

Substituting values for the physical parameters gives

$$r_i = 0.029 \sqrt{\frac{K \Delta t}{\phi \mu c_T}}$$

$$= 0.029 \sqrt{\frac{250 \times \Delta t}{.228 \times 0.71 \times 13 \times 10^{-6}}}$$

$$= 316 \sqrt{\Delta t}$$

Table 13-5 shows values of r_i for shut-in times of 0.25 day, 0.5 day and 1 day.

Table 13-5

Estimating the Radius of Investigation

Shut-in time		Radius of Investigation
days	hrs	[ft]
0.25	6	770
0.50	12	1100
1.00	24	1550

The DST showed that a no-flow boundary exists within approximately 700 ft of production well P-1.

13.5 Fluid Properties

In addition to pressure, flow capacity and boundary information, the DST was used to acquire a fluid sample. Table 13-6 presents fluid properties from a laboratory analysis of the DST fluid sample.

Table 13-6

Fluid Properties

	Oil			Gas		Water	
Pressure	Vis	FVF	Rso	Vis	FVF	Vis	FVF
psia	cp	RB/ STB	SCF/ STB	cp	RCF/ SCF	cp	RB/ STB
14.7	1.04	1.06	1	0	0.9358	0.5	1.019
514.7	0.910	1.207	150	0.0112	0.0352	0.5005	1.0175
1014.7	0.830	1.295	280	0.0140	0.0180	0.5010	1.0160
1514.7	0.765	1.365	390	0.0165	0.0120	0.5015	1.0145
2014.7	0.695	1.435	480	0.0189	0.0091	0.5020	1.0130
2514.7	0.641	1.500	550	0.0208	0.0074	0.5025	1.0115
3014.7	0.594	1.550	620	0.0228	0.0063	0.5030	1.0100
4014.7	0.510	1.600	690	0.0260	0.0049	0.5040	1.0070
5014.7	0.450	1.620	730	0.0285	0.0040	0.5050	1.0040
6014.7	0.410	1.630	760	0.0300	0.0034	0.5060	1.0010

Initial reservoir pressure was estimated from the DST to be 3936 psia at a depth of 9360 ft below sea level. This pressure is over 1400 psia greater than the laboratory measured bubble point pressure of 2514 psia. Table 13-6 presents fluid properties for undersaturated oil that must be corrected for use in a reservoir simulator (Chapter 5).

13.5.1 Black Oil PVT Correction

The corrections for adjusting laboratory measured differential liberation and separator data to a form suitable for use in a black oil simulator are given by the conversion equations [Moses, 1986]:

$$B_o(p) = B_{od}(p) \frac{B_{ofbp}}{B_{odbp}}$$

$$R_{so}(p) = R_{sofbp} - \left[R_{sodbp} - R_{sod}(p) \right] \frac{B_{ofbp}}{B_{odbp}}$$

where B_o is the oil formation volume factor and R_{so} is the solution gas-oil ratio. The subscripts are defined as d = differential liberation data; f = flash data; and bp = bubble point. For the case study, laboratory measurements include a flash from 6000 psig to 0 psig. This gives the following results:

B_{ofbp}	1.50 RB/STB	R_{sodbp}	760 SCF/STB
B_{odbp}	1.63 RB/STB	R_{sofbp}	650 SCF/STB

The corresponding correction factors are

$$B_o(p) = B_{od}(p) \times \frac{1.50}{1.63} = B_{od}(p) \times 0.92$$

$$R_{so}(p) = 650 - \left[760 - R_{sod}(p) \right] \times 0.92$$

Applying these corrections to the differential data yields the corrected results shown in Tables 13-7a, b, and c.

Table 13-7a
Corrected Oil Phase Properties

Pressure (psia)	Oil Vis (cp)	Oil FVF (RB/ STB)	Oil Rso (SCF/ STB)	Gas Vis (cp)	Gas FVF (RCF/SCF)	Water Vis (cp)	Water FVF (RB/STB)
14.7	1.040	1.062	1	0.0080	0.9358	0.5000	1.0190
514.7	0.910	1.207	150	0.0112	0.0352	0.5005	1.0175
1014.7	0.830	1.295	280	0.0140	0.0180	0.5010	1.0160
1514.7	0.765	1.365	390	0.0165	0.0120	0.5015	1.0145
2014.7	0.695	1.435	480	0.0189	0.0091	0.5020	1.0130
2514.7	0.641	1.500	550	0.0208	0.0074	0.5025	1.0115
3014.7	0.594	1.550	620	0.0228	0.0063	0.5030	1.0100
4014.7	0.510	1.600	690	0.0260	0.0049	0.5040	1.0070
5014.7	0.450	1.620	730	0.0285	0.0040	0.5050	1.0040
6014.7	0.410	1.630	760	0.0300	0.0034	0.5060	1.0010

Table 13-7b
Corrected Oil Phase Properties

Separator Test (Flash)		
Sep. P (psig)	GOR (SCF/STB)	FVF (RB/STB)
100	572	
↓		
0	78	
Total GOR = 650		1.5

Table 13-7c
Corrected Oil Phase Properties

Pressure (psia)	Oil FVF (RB/STB)	Oil Rso (SCF/STB)
14.7	1.062	1
514.7	1.110	89
1014.7	1.191	208
1514.7	1.256	310
2014.7	1.320	392
2514.7	1.380	457
3014.7	1.426	521
4014.7	1.472	586
5014.7	1.490	622
6014.7	1.500	650

13.5.2 Undersaturated Oil Properties

Slopes for undersaturated oil properties (Chapter 19.6) are calculated from Table 13-8.

Table 13-8
Undersaturated Oil Properties

Pressure (psia)	Corrected B_{opb} (RB/STB)	μ_o (cp)	Remarks
2515	1.3800	0.641	Bubble Point
3935	1.3473	0.706	Undersaturated Values

The rate of change of oil FVF with respect to pressure for the undersaturated oil is approximated by the difference

$$\frac{\Delta B_o}{\Delta P} = \frac{1.3473 - 1.3800}{3935 - 2515} \approx -0.000023 \frac{RB/STB}{psia}$$

This linear approximation is reasonable in many cases. Nonlinear, undersaturated slopes can be modeled by some simulators.

Similarly, the rate of change of oil viscosity with respect to pressure for the undersaturated oil is

$$\frac{\Delta \mu_o}{\Delta P} = \frac{0.706 - 0.641}{3935 - 2515} \approx 0.000046 \frac{RB/STB}{psia}$$

The rate of change of solution GOR (R_{so}) is zero in the pressure regime above the bubble point pressure.

13.6 Reservoir Management Constraints

Reservoir management constraints are presented in Table 13-9. They include limits on capital expenditures, such as the number of wells that can be drilled, and operating constraints. In this case, for example, it is considered important to keep water-oil ratio (WOR) less than five STB water/STB oil. These constraints are typically imposed by considerations ranging from technical to commercial. The constraints are especially important in the prediction phase of the study.

Table 13-9
Reservoir Management Constraints

♦ One additional well may be drilled.
♦ Completion interval in existing well may be changed.
  ◇ The well is presently completed in entire pay interval.
♦ Target oil rate ≈ 1000 STB/D
♦ Water is available for injection if desired.
♦ Limit WOR < 5
♦ Minimum allowed BHP = 2600 psia
♦ Maximum allowed injection pressure = 5000 psia
♦ Minimum economic oil rate = 100 STB/D

Exercises

Exercise 13.1 Plot FVF, viscosity, and solution GOR versus pressure for saturated and undersaturated oil.

Exercise 13.2 Verify that the PVT values are properly entered in data set CS-MB.DAT. What is the bubble point pressure in the model?

References

Clark, N.J. (1969): **Elements of Petroleum Reservoirs**, Richardson, TX: Society of Petroleum Engineers.

Moses, P.L. (July 1986): "Engineering Applications of Phase Behavior of Crude Oil and Condensate Systems," *Journal of Petroleum Technology*, pp. 715-723; and F.H. Poettmann and R.S. Thompson (1986): "Discussion of Engineering Applications of Phase Behavior of Crude Oil and Condensate Systems," *Journal of Petroleum Technology*, pp. 1263-1264.

Chapter 14

Volumetrics and Material Balance

14.1 Volumetrics

A volumetric estimate of oil volume is a useful number for checking the accuracy of the numerical representation of the reservoir in a reservoir model. The volume of oil in the reservoir may be estimated as the product of bulk volume V_B, porosity ϕ, and oil saturation S_o.

The bulk volume of the reservoir is estimated by writing bulk volume V_B as the product $\Delta x \Delta y \Delta z$ where Δx, Δy, and Δz approximate the length, width, and net thickness of the pay interval, respectively.

♦ From maps: $\Delta x = 2000'$ and $\Delta y = 1200'$ for an area ≈ 55 acres
♦ From well logs (Chapter 13): $\Delta z = 72' + 96' = 168'$

The resulting estimate of bulk volume V_B is 4.03×10^8 ft^3.

Pore Volume V_P is the product ϕV_B. Porosity ϕ is estimated as the thickness weighted average porosity from well logs:

$$\phi \approx \frac{72 \times 0.20 + 96 \times 0.25}{168} \approx 0.228$$

Taking the product of porosity and bulk volume gives an estimate of the pore volume:

$$V_P = \phi V_B \approx 9.18 \times 10^7 \text{ft}^3 \approx 16.4 \times 10^6 RB$$

The product of oil saturation and pore volume gives an estimate of oil volume in reservoir barrels. Dividing this volume by an average oil formation volume

141

factor B_o for the reservoir gives an estimate of oil volume in stock tank barrels. The value of oil FVF at an initial average reservoir pressure of 3935 psia is 1.3473 RB/STB. This value is obtained from laboratory data that has been corrected for use in a reservoir simulator (Chapter 13.5). The resulting oil volume is

$$V_o = \frac{S_o V_P}{B_o} \approx \frac{0.7\, V_P}{B_o} \sim \frac{11.5 \times 10^6\, \text{RB}}{1.3473\, \text{RB/STB}} \sim 8.5 \times 10^6\, \text{STB}$$

14.2 Material Balance

Volumetrics provides one measure of the quality of a reservoir model, but it is based on information that does not change with time. Another estimate of original oil volume can be obtained from a material balance study if a reasonable amount of production data is available, such as the historical data presented in Chapter 13. At this point we have surmised that the reservoir was initially undersaturated, but it may not have aquifer support.

The presence of a few barrels of water during the latter months of the first year of production indicates that mobile water is present, but its source is unknown. The volume of produced water is small enough to be water mobilized by swelling as reservoir pressure declines, or it could be the first indication of water breakthrough from aquifer influx. Both of these scenarios can be assessed if we consider the possibilities of depletion with and without aquifer influx.

We begin by deriving the material balance equation for the more general case: depletion of an undersaturated oil reservoir with water influx. The derivation is simplified by assuming formation compressibility is negligible and then setting the decrease in oil volume at reservoir conditions equal to the increase in water volume at reservoir conditions as oil is produced and reservoir pressure decreases.

1. Calculate the decrease in oil volume ΔV_o (RB):

Let N = original oil in place = OOIP (STB)

   B_{oi} = oil FVF (RB/STB) at initial pressure P_i

N_p = oil produced (STB) at pressure P and time t

B_o = oil FVF (RB/STB) at pressure P and time t

Then NB_{oi} = OOIP (RB) at initial reservoir pressure P_i

$\quad (N - N_p) B_o$ = OIP (RB) at pressure P and time t

Therefore the change in oil volume is given by

$$\Delta V_o = NB_{oi} - (N - N_p) B_o$$

2. Calculate the increase in water volume ΔV_w (RB):

Let $\quad W$ = original water in place = OWIP (RB) at initial pressure P_i

$\quad\quad B_w$ = water FVF (RB/STB) at pressure P and time t

$\quad\quad W_p$ = water produced (STB) at pressure P and time t

$\quad\quad W_e$ = water influx (RB)

Then $W_p B_w$ = cumulative water produced (RB) at pressure P and time t

Therefore the change in water volume is given by

$$\Delta V_w = (W + W_e - W_p B_w) - W = W_e - W_p B_w$$

3. The assumption that the volume of the reservoir remains constant implies $\Delta V_o = \Delta V_w$. Combining results from steps 1 and 2 above gives the material balance equation for depletion of an incompressible, undersaturated oil reservoir with aquifer influx:

$$NB_{oi} - (N - N_p) B_o = W_e - W_p B_w$$

The two unknowns in the equation are N and W_e.

The simplest production scenario is to assume that water influx is negligible, that is, $W_e = 0$. If we further observe that water production W_p is insignificant, we have

$$N = \frac{N_p B_o}{B_o - B_{oi}}$$

where $B_{oi} = 1.3473$ RB/STB at $P_i = 3935$ psia. Oil FVF has been corrected for use in this calculation (see Chapter 13 for details). The corresponding estimate for OOIP is $N \approx 1500\ N_p$ with $B_o - B_{oi} \approx 0.0009$ RB/STB. The results of the calculation follow:

have class calculate knowing from 3935 - 25/5 B_o slope
have class calculate

Table 14-1
Results Assuming No Water Influx

Time (days)	Pressure (psia)	B_o (RB/STB)	N_p (MSTB)	N (MMSTB)
91	3935	1.3482	46	69
183	3898	1.3482	91	136
274	3897	1.3482	137	205
365	3895	1.3483	183	274

The value of N increases at each time. This implies that the material balance model does not account for all of the pressure support and suggests that an aquifer influx model should be considered.

If we use a volumetric estimate of N, namely $N_{vol} = 8.5$ MMSTB from Chapter 14.1, we can calculate W_e. Again recognizing that $W_p \approx 0$, the material balance equation becomes

$$W_e = N(B_{oi} - B_o) + N_p B_o$$

Results of the calculation are shown in Table 14-2.   *We?*

Table 14-2
Results Assuming Water Influx with Volumetric OOIP

Time (days)	Pressure (psia)	B_o (RB/STB)	N_p (MSTB)	N (MMSTB)
90	3935	1.3482	46	54 (52)
180	3898	1.3482	91	115 (113)
270	3897	1.3482	137	177 (174)
365	3895	1.3483	183	239 (234)

Notice that W_e increases as a function of time. The values in parentheses are BOAST4D values when the correct aquifer model is used.

Exercises

Exercise 14.1 Verify the calculations reported in Tables 14-1 and 14-2.

Exercise 14.2 Data file CS-MB.DAT is an input file for a material balance analysis of the case study. It represents the reservoir as a single grid block, or "tank" model. The tank model is equivalent to a material balance calculation. Run BOAST4D with the file CS-MB.DAT. Verify that the original volume of oil in the model agrees with the volumetric estimate presented in Chapter 14.1.

Exercise 14.3 Use data file CS-MB.DAT to study the effect of aquifer influx on material balance performance. This is done by modifying the input data set to include an aquifer model, then adjusting aquifer parameters until model pore volume weighted average reservoir pressures match the pressures reported in Chapter 13.3. Note: the pore volume weighted average reservoir pressure P_{av} is given by

$$P_{av} = \frac{\sum\limits_{j=i}^{N} P_j V_{Pj}}{\sum\limits_{j=i}^{N} V_{Pj}}$$

where N is the total number of grid blocks in the model grid, P_j is the oil phase pressure in grid block j, and V_{pj} is the pore volume of grid block j. Chapter 19.10 contains details on how to set up an analytic aquifer. For an example of a data set with an analytic aquifer model, see data file EXAM9.DAT.

Chapter 15

Model Initialization

Chapter 14 presented basic volumetric and material balance analyses using data provided in Chapter 13. Exercise 14.3 began the history matching process by attempting to identify the impact of aquifer influx on model performance. As we continue our preparation of a three-dimensional simulation model, we observe that not all of the data needed by the simulator is available. Since we cannot ignore data and still perform a credible model study, we must complete the data set. Several options are available, such as ordering additional measurements or finding reasonable correlations or analogies for the missing data. In this case, our commercial interests are best served by moving the project forward without additional expense or delays.

15.1 Relative Permeability

We do not have laboratory measured relative permeability data. We could attempt to construct relative permeability data from production data, but our production history is essentially single phase oil. Since we must specify relative permeability to run the model, we can turn to analogous reservoirs or correlations for guidance. Let us choose the Honarpour, et al. [1982] correlation for a water-wet sandstone as a starting point for determining relative permeability curves. Well logs provide some information about saturation end points such as initial and irreducible water saturation. Core floods and capillary pressure measurements could provide information about residual hydrocarbon saturations, but they are not available. For that reason, end points like residual oil saturation must be estimated. Results of the calculation are shown in BOAST4D format (Chapter 19.5) in Table 15-1 and Figure 15-1. The acronyms in Table 15-1 are defined as follows:

- ◆ SAT is the saturation associated with each phase
- ◆ KROW is the relative permeability of oil in the presence of water expressed as a function of oil saturation
- ◆ KRW is the relative permeability of water in a water-oil system expressed as a function of water saturation
- ◆ KRG is the relative permeability of gas in a gas-oil system expressed as a function of gas saturation
- ◆ KROG is the relative permeability of oil in the presence of gas expressed as a function of oil saturation

Table 15-1
Relative Permeability

SAT	KROW	KRW	KRG	KROG
0.000	0.000	0.000	0.000	0.000
0.030	0.000	0.000	0.000	0.000
0.050	0.000	0.000	0.020	0.000
0.100	0.000	0.000	0.090	0.000
0.150	0.000	0.000	0.160	0.000
0.200	0.000	0.000	0.240	0.000
0.250	0.000	0.000	0.330	0.000
0.300	0.000	0.000	0.430	0.000
0.350	0.001	0.005	0.550	0.000
0.400	0.010	0.010	0.670	0.000
0.450	0.030	0.017	0.810	0.000
0.500	0.080	0.023	1.000	0.000
0.550	0.180	0.034	1.000	0.000
0.600	0.320	0.045	1.000	0.000
0.650	0.590	0.064	1.000	0.000
0.700	1.000	0.083	1.000	0.000
0.800	1.000	0.120	1.000	0.000
0.900	1.000	0.120	1.000	0.000
1.000	1.000	0.120	1.000	0.000

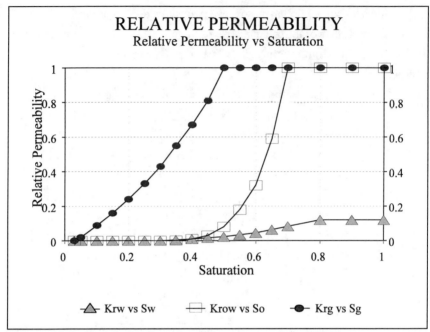

Figure 15-1. Correlation based on Honarpour, et al. [1982].

If our choice of correlations does not match field performance, we will have to change the relative permeability curves. In any event, we recognize that in this case study relative permeability is poorly known and should be considered highly uncertain.

15.2 Fluid Contacts

A water-oil contact (WOC) was not seen on either well logs or seismic data. The production of a small amount of water suggests that there may be a WOC in the vicinity of the reservoir. The data are not compelling however. We could assume the oil zone extends well below the bottom depth of our well, but this would be an optimistic assumption that could prove to be economically disastrous. In the interest of protecting our investment, let us make the more conservative assumption that a WOC does exist and is just beyond the range of our observations, namely well log and seismic data. We assume WOC ≈ 9600 ft, which is near the bottom of the seismically observed reservoir structure. The pressure at this WOC depth is estimated to be about 4000 psia.

15.3 Grid Preparation

Figure 15-2 is a sketch of the well location relative to the interpreted reservoir boundaries. Based on seismic data shown in Chapter 1.2, the reservoir is thought to be bounded to the east by a facies change.

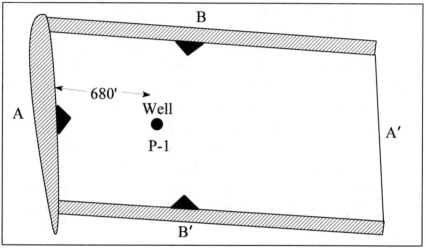

Figure 15-2. Plan view.

A cross-section through points B and B′ is shown in Figure 15-3. The sides of the reservoir appear to be bounded by faults. Without evidence to the contrary, we assume that the faults are sealing. This assumption is subject to verification during the history match phase of the study.

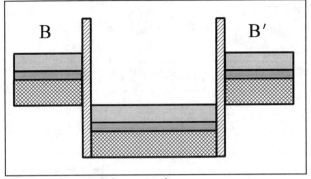

Figure 15-3. BB′ Cross-section.

A cross-section through points A and A′ is sketched in Figure 15-4. It illustrates the dip of the reservoir and the layering. The structure of the reservoir is based on well log and seismic interpretation. The downdip fault is speculative. It is based on the assumption that the fault shown on the western side of Figure 15-2 extends down through the formation. This is not obvious from seismic data. Indeed, if the reservoir is receiving aquifer support, the aquifer influx will come from downdip as the reservoir is depleted. Bear in mind, however, that both the fault and the aquifer may be present. This could happen, for example, if the fault is not sealing. The fault could be providing a flow path for water influx from another horizon.

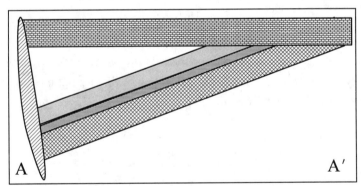

Figure 15-4. AA′ Cross-section.

Exercises

Exercise 15.1 Data set CS-VC.DAT is a vertical column model of the case study. Sketch the grid to scale, locate the contacts on the sketch, and match reservoir pressure.

Exercise 15.2 Repeat Exercise 15.1 beginning with the cross-section model data set CS-XS.DAT.

References

Honarpour, M., L.F. Koederitz, and A.H. Harvey (1982): "Empirical Equations for Estimating Two-Phase Relative Permeability in Consolidated Rock," *Journal of Petroleum Technology*, pp. 2905-2908.

Chapter 16

Well Model Preparation

The history match is now well under way. The models discussed in the exercises of Chapter 15 are conceptual models designed to provide you with a sense of how fluids move in the reservoir. This is the art of modeling. As you work with various models of the reservoir, you should begin to develop a knowledge base for determining how changes to model parameters will help achieve a match for a particular observable. This knowledge base is valuable as you develop your feel for the study.

At this point in the study, we are ready to work with the full three-dimensional model. First, however, it is pedagogically wise to review the well model.

16.1 Productivity Index Estimate

Well model calculations in BOAST4D need to have the quasi-stationary productivity index factor (PID) specified by the user. PID is estimated from the expression (Chapter 29)

$$\text{PID} = \frac{0.00708\,K_{abs}\,h_{net}}{\ln(r_e/r_w) + S}$$

where

r_e	=	drainage radius (ft)
r_w	=	wellbore radius (ft)
S	=	skin
K_e	=	$k_{ro}K_{abs}$ = effective permeability (md)
h_{net}	=	net thickness (ft)

Given $S = -0.5$, $r_w = 0.25$ ft and

$$r_o \simeq 0.14(\Delta x^2 + \Delta y^2)^{\frac{1}{2}} \simeq 40 \text{ ft}$$

with $\Delta x = \Delta y = 200$ ft., we find

$$PID = 1.55 \times 10^{-3} K_{abs} h_{net}$$

where $r_e \simeq r_o$. Table 16-1 presents the calculation of *PID* for each layer identified by well log analysis.

Table 16-1

Estimate of PID by Layer

Layer	K_{abs} [md]	h_{net} [ft]	PID
1	75	72	8.4
2	0	20	0
3	250	64	24.8
4	250	32	12.4

16.2 Oil Well FBHP Estimate

The production well model needs a flowing bottomhole pressure (FBHP). Assuming an oil column in the wellbore, we can prepare a quick estimate of FBHP for a single-phase oil well that is completed at a 9500 ft depth by assuming FBHP ≈ oil head. Consequently, oil head is approximated by

$$\gamma_o \Delta z \approx \text{FBHP}$$

where γ_o is the oil pressure gradient and Δz is the height of the oil column. An estimate of average oil pressure gradient for the oil column is found by averaging the pressure gradient at surface and reservoir conditions:

♦ Approximate pressure gradient at surface conditions:

$$\rho_s = 46.244 \ \frac{lb}{ft^3} \approx 0.321 \ \frac{psia}{ft}$$

where oil density at surface conditions (ρ_s) is 46.244 lbm/SCF.

♦ Approximate pressure gradient at reservoir conditions:

$$\rho_R = \frac{\rho_s}{B_o} \approx 34.3 \ \frac{lb}{ft^3} \approx 0.238 \ \frac{psia}{ft}$$

where oil FVF (B_o) at bottomhole conditions is 1.3482 RB/STB.

The resulting FBHP for use in BOAST4D is

$$FBHP = \frac{1}{2} \left[0.321 \ \frac{psia}{ft} + 0.238 \ \frac{psia}{ft} \right] \times 9500 \, ft \approx 2660 \, psia$$

A more accurate estimate can be obtained from wellbore correlations or nodal analysis as discussed by such authors as Brown and Lea [1985].

16.3 Well Block Pressure from PBU

In Chapter 9 we saw that a pressure correction was needed to properly relate the pressure buildup (PBU) curve to simulator well block pressures. To illustrate this correction, suppose a well is in a block with grid dimensions Δx = 200 ft and Δy = 200 ft. We want to compare the simulator well block pressure with a pressure from a PBU. Peaceman [1978, 1983] showed that shut-in pressure P_{ws} of the actual well should equal the simulator well block pressure P_o at a shut-in time Δt_s given by

$$\Delta t_s = \frac{1688 \, \phi \, \mu \, c_T \, r_o^2}{K}$$

For an isotropic reservoir in which horizontal permeability does not depend on direction, that is, $K_x = K_y$, we estimate the equivalent radius of a well in the center of a grid block as

$$r_o \approx 0.14 (\Delta x^2 + \Delta y^2)^{\frac{1}{2}}$$

The shut-in time Δt_s at which the PBU pressure should be obtained are calculated from the following physical parameters:

c_r	3×10^{-6} psia^{-1}
c_o	13×10^{-6} psia^{-1}
c_w	3×10^{-6} psia^{-1}
S_o	0.7
S_w	0.3
μ_o	0.71 cp
ϕ	0.20
K	75 md

The equivalent radius of the well block is estimated to be $r_o \approx 0.14 \, (200^2 + 200^2)^{\frac{1}{2}} = 39.6$ ft, while the total compressibility is given by $c_T = c_r + S_o \, c_o + S_w \, c_w = 3 \times 10^{-6} + 0.7 \, (13 \times 10^{-6}) + 0.3 \, (3 \times 10^{-6}) \approx 13 \times 10^{-6}$ psia^{-1}. The PBU shut-in time corresponding to these values is

$$\Delta t_s = 1688 \, \frac{(0.20) \, (0.71) \, (13 \times 10^{-6}) \, (39.6)^2}{75}$$

$$= 0.065 \text{ hr.} \approx 4 \text{ minutes}$$

This early time part of the PBU curve could be masked by wellbore storage effects. Since the shut-in pressure P_{ws} of the actual well equals the simulator well block pressure P_o at a shut-in time Δt_s, the shut-in pressure P_{ws} may have to be obtained by extrapolation of the radial flow curve.

16.4 Throughput Estimate

Model time step size is estimated by calculating pore volume throughput:

$$V_{PT} = \frac{Q \Delta t}{V_P} \quad (5.6146)$$

where $V_P = \phi \, \Delta x \, \Delta y \, \Delta z$ = Pore Volume (ft^3)

Q = Volumetric flow rate at reservoir conditions (RB/day)

Δt = time step size (day)

Time steps for an IMPES simulator should correspond to about 10% throughput or less. The maximum time step that should be used can be estimated as follows.

Suppose $\phi = 22.5\%$, $\Delta x = \Delta y = 200'$, $\Delta z = h_{net}$, and $Q = 400$ RB/day. Then Δt is found by setting $V_{PT} = 0.10$ and rearranging the pore volume throughput equation to give

$$\Delta t = \frac{(0.1)V_P}{5.6146\,Q} = (0.1)\frac{\phi\,\Delta x\,\Delta y\,\Delta z}{5.6146\,Q}$$

$$= 0.4\,h_{net} \ \text{(days)}$$

If $h_{net} = 100$ ft, then $\Delta t \approx 40$ days is an estimate of the maximum IMPES time step size.

Exercises

Exercise 16.1 Repeat the shut-in time calculation using $\Delta x = 1000$ ft and $\Delta y = 1000$ ft. The new shut-in time Δt_s should be less than one hour.

Exercise 16.2 Run data set CS-XS.DAT with maximum time step sizes ranging from 15 days to 60 days. Select a maximum time step size by monitoring the material balance error and the stability of the solution. A solution is unstable if it oscillates, that is, variables like GOR or WOR vary between a high and low value from one time step to the next.

Exercise 16.3 What is the effect of doubling the PID in data set CS-XS.DAT?

Exercise 16.4 How does model performance change if skin $S = 0$?

Exercise 16.5 What is the effect of reducing the well FBHP by 1000 psia? The reduction in FBHP is one way to simulate gas lift or pumping.

References

Brown, K.E. and J.F. Lea (Oct. 1985): "Nodal Systems Analysis of Oil and Gas Wells," *Journal of Petroleum Technology*, pp.1751-1763.

Peaceman, D.W. (June 1978): "Interpretation of Well-Block Pressures in Numerical Reservoir Simulation," *Society of Petroleum Engineering Journal*, pp. 183-194.

Peaceman, D.W. (June 1983): "Interpretation of Well-Block Pressures in Numerical Reservoir Simulation with Nonsquare Grid Blocks and Anisotropic Permeability," *Society of Petroleum Engineering Journal*, pp. 531-543.

Chapter 17

History Matching and Predictions

The previous chapters set the stage for preparing a three-dimensional model of the case study reservoir. A three-dimensional model should provide enough reservoir definition to let us make meaningful performance predictions. The goal of this chapter is not to lecture, but to coach you through a series of exercises. We begin with a review of the three-dimensional history match model.

17.1 Full Field (3D) Model History Match

Data file CS-HM.DAT is the three-dimensional model used to prepare the production history presented in Chapter 13.3. The grid in Figure 17-1 was used to model the reservoir shown in Figure 15-2. Each grid block is a square with lengths $\Delta x = \Delta y = 200$ ft. The dark areas of the grid are outside of the reservoir area. The pore volume in the dark area is made inactive in data file CS-HM.DAT by using porosity multipliers.

The depth and thickness of each grid block depend on reservoir architecture. The model grid should approximate the structure depicted in Figure 15-4, which is based on Figures 13-1 and 13-2. The dip of the reservoir is included by specifying the tops of each grid block. The grid block length modifications are designed to cut off those parts of the block that continue the grid beyond the surface of the unconformity sketched in Figure 15-4.

Transmissibility multipliers in the vertical direction are set to 0 to simulate impermeable shale barriers. This includes the shale streak that divides the second major sand into two thinner sands with a shale break. The interpretation of seismic data was unable to resolve this feature, but the well log shown in Figure 17-2 does indicate the presence of a shale streak.

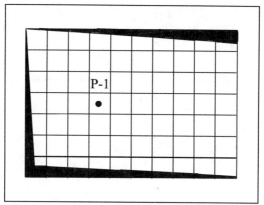

Figure 17-1. Plan view of grid.

The water-oil contact is at 9600 ft. A steady state aquifer is in communication with all three oil layers at this depth. It is the source of water production shown in Table 13-2.

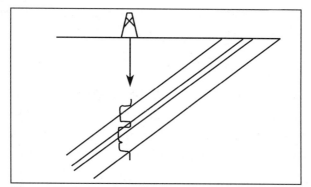

Figure 17-2. Overlay of seismic and well log data.

17.2 Predictions

Now that we have a history match model, we are ready to make predictions. The first step is to establish a base case prediction which assumes there will be no changes in operating strategy. Given a base case prediction, several runs should be made to optimize reservoir performance within the constraints imposed by the commissioners of the study. If the model is run with well P-1 switched from oil rate to bottom hole pressure control, the PI for well P-1 needs to be calibrated to assure continuity in the oil rate. The following exercises are designed to guide you through the prediction process.

Exercises

Exercise 17.1 Data set CS-HM.DAT was used as the basis of the case study. Run data set CS-HM.DAT and verify that it matches the data shown in Table 13-2.

Exercise 17.2 Several sensitivity runs may be made by varying model parameters and noting reservoir performance. As an example of a sensitivity study, vary the WOC by ±100 ft. How does this variation affect water breakthrough and oil recovery during the history match period?

Exercise 17.3 Run data set CS-HM.DAT for five years (four years into the future) with Well P-1 under oil rate control. This run establishes a base case prediction.

Exercise 17.4 Data set CS-PD.DAT is the base case prediction. Beginning with this data set, maximize oil recovery given the constraints listed in Table 13-9. Two ideas to consider are downdip water injection after drilling an updip producer; and downdip production after drilling an updip gas injector.

Part III

BOAST4D

User's Manual

Chapter 18

Introduction to BOAST4D

BOAST4D simulates isothermal, Darcy flow in up to three dimensions. It assumes reservoir fluids can be described by up to three fluid phases (oil, gas, and water) with physical properties that depend on pressure only. Gas is allowed to dissolve in both the oil and water phases. A feature unique to BOAST4D is the inclusion of compressional velocity and acoustic impedance calculations. These reservoir geophysical calculations make it possible to track changes in seismic variables as a function of time, which is the basis for 4D seismic analysis.

BOAST4D was designed to run on DOS-based personal computers with 486 or better math co-processors. This size simulator is well-suited for learning how to use a reservoir simulator, developing an understanding of reservoir management concepts, and for solving many types of reservoir engineering problems. It is an inexpensive tool for performing studies that call for more sophistication than is provided by analytical solutions, yet do not require the use of full-featured commercial simulators.

BOAST4D is a modified version of the black oil simulator BOAST II that was published by the U.S. Department of Energy in 1987 [Fanchi, et al., 1987]. BOAST II was an improved version of BOAST, an implicit pressure-explicit saturation (IMPES) simulator published by the U.S. Department of Energy in 1982 [Fanchi, et al., 1982]. There have been several modifications of BOAST II published by the Bartlesville Project Office of the U.S. Department of Energy. BOAST4D is based on BOAST II.

A comparison of differences between BOAST II and BOAST4D is given in the following tables. The first table shows that a variety of useful geophysical and reservoir engineering features have been added to BOAST4D, including the ability to perform material balance studies with a tank model, the representation of horizontal or deviated wells, and the calculation of important reservoir geophysical information.

Table 18-1
Comparison of Reservoir Modeling Differences

FEATURE	BOAST II	BOAST4D
Material balance tank model (1 grid block)	Not available	New
Well completions	Vertically contiguous	Flexible - may skip layers
Horizontal well	Not available	New
Slanted well	Not available	New
Compressional velocity	Not available	New
Shear velocity	Not available	New
Acoustic impedance	Not available	New
Reflection coefficient	Not available	New
Modify ϕ, K	Input ϕ, K	Added multiply by factor
Modify transmissibility	Input transmissibility	Multiply trans. by factor
Saturation initialization	User specified	Added gravity segregated option

In addition, BOAST4D includes changes to improve computational performance, that is, a more accurate algorithm for interpolating gas formation volume factor B_g.

Table 18-2
Comparison of Computational Differences

FEATURE	BOAST II	BOAST4D
Interpolation	B_g	$b_g = 1/B_g$ Improves material balance
Saturation table end points	Set to -0.1 and 1.1	Set to 0.0 and 1.0
Time stepping and reports	Counter and user specified	Simplify to user specified only
Debug codes	Optional	Deleted
Restart	Available	Deleted - restart by specifying arrays
Stabilized IMPES	Available	Deleted - not robust

BOAST II has been tested under a wide range of conditions. Detailed comparisons with other simulators were made for four types of problems: oil and gas depletion, waterflooding, gas injection with constant bubble point pressure, and gas injection with variable bubble point pressure. Favorable comparisons were observed with respect to oil rates, GORs, gas saturations, and pressures. The one exception is the reservoir pressure comparison for the variable bubble point pressure case. In this case, BOAST II reservoir pressures were consistently lower than other simulator values. A mass conserving expansion of accumulation terms can improve the accuracy for variable bubble point pressure problems [Fanchi, 1986], but the mass conserving expansion option requires additional run time and is not included in BOAST4D.

BOAST4D retains the robustness of BOAST II while substantially increasing program accuracy. BOAST4D has an improved interpolation algorithm that reduces material balance error for some problems by as much as a factor of ten relative to the DOE versions BOAST and BOAST II. This feature increases the range of applicability of BOAST4D and is especially valuable for gas and gas-oil systems. The algorithm does not degrade program speed.

18.1 Program Configuration

The user needs to have at least 575 kB RAM to run BOAST4D with the following standard configuration.

Table 18.3
Standard Configuration of BOAST4D

Maximum number of blocks in x direction	10
Maximum number of blocks in y direction	10
Maximum number of blocks in z direction	4
Maximum number of Rock regions	3
Maximum number of entries in a Rock region table	30
Maximum number of PVT regions	3
Maximum number of entries in a PVT region table	30
Maximum number of wells	25
Maximum number of connections per well	5
Total number of blocks using 1D direct matrix solution methods	10
Total number of blocks using 2D or 3D direct matrix solution methods	400

Although BOAST4D runs satisfactorily from a disk, program performance can be substantially enhanced by copying the disk files to the hard drive. The following procedure is recommended for a disk drive A and hard drive C:

- ♦ Make a directory, e.g. type MD C:\BOAST4D\
- ♦ Connect to the directory, e.g. type CD C:\BOAST4D\
- ♦ Copy files from the disk, e.g. type COPY A:*.*
- ♦ Run BOAST4D, e.g. type B4D
- ♦ Respond to questions, e.g. choose output option 6 to write BOAST4D output to the terminal screen

18.2 Input Data File - BTEMP.DAT

BOAST4D reads a file called BTEMP.DAT and outputs to files BTEMP.TSS, BTEMP.PLT, BTEMP.WEL and BTEMP.OUT. The output files are described in Chapter 21. You should use the command RENAME to rename any runs you wish to save because BOAST4D overwrites the BTEMP.* files.

The easiest way to prepare a new data file is to edit an old one. This will give you an example of the formats needed for most options. If you start with an old data set, make sure that you check all applicable data entries and make changes where appropriate.

18.3 Data Input Requirements

BOAST4D input data is divided into two parts: initialization data, and recurrent data. Initialization data is described in Chapter 19. It includes data that is set at the beginning of the study and is not expected to change during a model run. Such data includes the reservoir description and fluid properties. Recurrent data is described in Chapter 20 and refers to data that is expected to change during the course of a simulation. It includes well schedules and time step control information. Additional discussion of BOAST4D is presented in Part IV: Technical Supplement.

Title or heading records are read before each major and many minor sections. These records are designed to make the input data file easier to read and edit.

All input data, with the exception of well names, is entered by free format. Data entered on the same line must be separated by a comma or a space.

In many cases, codes are read that will specify the type of input to follow and the number of values that will be read. These codes increase the efficiency and flexibility of entering input data.

Input tabular data should cover the entire range of values expected to occur in a simulation. The linear table interpolation algorithms in BOAST4D will return tabluated endpoint values if the independent variable goes outside the range of the input tabular values. No message will be printed if this occurs.

If an array of input values must be read, the following input order must be followed. Layer 1 ($K = 1$) is read first. The data in each layer are read by rows, starting with row 1 ($J = 1$). Values of the array element are read for the first row starting with column 1 ($I = 1$) and proceeding to the end of the row (column $I = II$). After II values are read, the next row ($J = 2$) of values are entered. These values must begin on a new line. This data entry procedure is repeated for all rows and, subsequently, for all layers until the complete set of array elements has been entered. The data sets listed in Chapter 22 illustrate the data entry procedure.

References

Fanchi, J.R., K.J. Harpole, and S.W. Bujnowski (1982): "BOAST: A Three-Dimensional, Three-Phase Black Oil Applied Simulation Tool", 2 Volumes, U.S. Department of Energy, Bartlesville Energy Technology Center, OK.

Fanchi, J.R., J.E. Kennedy, and D.L. Dauben (1987): "BOAST II: A Three-Dimensional, Three-Phase Black Oil Applied Simulation Tool", U.S. Department of Energy, Bartlesville Energy Technology Center, OK.

Fanchi, J.R. (1986): "BOAST-DRC: Black Oil and Condensate Reservoir Simulation on an IBM-PC," SPE Paper 15297, *Proceedings from Symposium on Petroleum Industry Applications of Microcomputers of SPE*, Silver Creek, CO, June 18-20.

Chapter 19

Initialization Data

Initialization data records are read once at the beginning of the simulation. They must be read in the order presented below.

1. **Title**   Up to 80 characters; this record will appear as run title.

19.1 Grid Dimensions and Geometry

19.1.1 Grid Dimensions

1. **Heading**   Up to 80 characters.

2. **II, JJ, KK**

Code	Meaning
II	number of grid blocks in the x direction
JJ	number of grid blocks in the y direction
KK	number of grid blocks in the z direction

3. **Heading**   Up to 80 characters.

4.   **KDX, KDY, KDZ, KDZNET**

KDX   Control code for input of x direction grid size.

KDY   Control code for input of y direction grid size.

KDZ   Control code for input of z direction gross grid block thicknesses.

KDZNET   Control code for input of z direction net grid block thicknesses.

Code	Value	Meaning
KDX	-1	The x direction grid dimensions are the same for all blocks in the grid. Read only one value.
	0	The x direction dimensions are read for each block in the first row (J = 1) of layer one (K = 1). These same values are assigned to all other rows and all other layers in the model grid. Read II values.
	1	The x direction dimensions are read for each block in layer one (K = 1). These same values are assigned to all other layers in the grid. Read II × JJ values.
KDY	-1	The y direction grid dimensions are the same for all blocks in the grid. Read only one value.
	0	The y direction dimensions are read for each block in the first column (I = 1) of layer one (K = 1). These values are assigned to all other columns and all other layers in the model grid. Read JJ values.
	1	The y direction dimensions are read for each block in layer one (K = 1). These same values are assigned to all other layers in the grid. Read II × JJ values.
KDZ	-1	The z direction gross thickness is the same for all blocks in the grid. Read only one value.
	0	A constant value of gross thickness is read for each layer in the grid; each layer may have a different, but constant value. Read KK values.

Code	Value	Meaning
KDZ	1	The z direction gross thickness is read for each block in the grid. Read II × JJ × KK values.
KDZNET	-1	The z direction net thickness is the same for all blocks in the grid. Read only one value.
	0	A constant value of net thickness is read for each layer in the grid; each layer may have a different, but constant value. Read KK values.
	1	The z direction net thickness is read for each block in the grid. Read II × JJ × KK values.

5. **DX**

    DX   Grid block size in x direction (ft).

    If KDX = -1, read one constant value.

    If KDX = 0, read II values (one for each row).

    If KDX = +1, read II × JJ values (one for each K = 1 block).

6. **DY**

    DY   Grid block size in y direction (ft).

    If KDY = -1, read one constant value.

    If KDY = 0, read JJ values (one for each column).

    If KDY = +1, read II × JJ values (one for each K = 1 block).

7. **DZ**

    DZ   Gross grid block thickness in z direction (ft).

    If KDZ = -1, read one constant value.

    If KDZ = 0, read KK values (one for each layer).

    If KDZ = +1, read II × JJ × KK values (one for each block).

8. **DZNET**

    DZNET   Net grid block thickness in z direction (ft).

    If KDZ = -1, read one constant value.

    If KDZ = 0, read KK values (one for each layer).

    If KDZ = +1, read II × JJ × KK values (one for each block).

19.1.2 Modifications to Grid Dimensions

1.   **Heading**   Up to 80 characters.

2.   **NUMDX, NUMDY, NUMDZ, NUMDZN, IDCODE**

    NUMDX   Number of regions where x direction grid size (DX) is changed.

    NUMDY   Number of regions where y direction grid size (DY) is changed.

    NUMDZ   Number of regions where z direction gross thickness (DZ) is changed.

    NUMDZN Number of regions where z direction net thickness (DZN) is changed.

    IDCODE   = 0 means do not print the modified distributions;

              = 1 means print the modified distributions.

3.   **I1, I2, J1, J2, K1, K2, DX**

    Omit this record if NUMDX = 0.

    I1   Coordinate of first region block in I direction.

    I2   Coordinate of last region block in I direction.

    J1   Coordinate of first region block in J direction.

    J2   Coordinate of last region block in J direction.

    K1   Coordinate of first region block in K direction.

    K2   Coordinate of last region block in K direction.

    DX   New value of x direction grid size for region (ft).

    NOTE: NUMDX records must be read.

4. **I1, I2, J1, J2, K1, K2, DY**

Omit this record if NUMDY = 0.

I1	Coordinate of first region block in I direction.
I2	Coordinate of last region block in I direction.
J1	Coordinate of first region block in J direction.
J2	Coordinate of last region block in J direction.
K1	Coordinate of first region block in K direction.
K2	Coordinate of last region block in K direction.
DY	New value of *y* direction grid size for region (ft).

NOTE: NUMDY records must be read.

5. **I1, I2, J1, J2, K1, K2, DZ**

Omit this record if NUMDZ = 0.

I1	Coordinate of first region block in I direction.
I2	Coordinate of last region block in I direction.
J1	Coordinate of first region block in J direction.
J2	Coordinate of last region block in J direction.
K1	Coordinate of first region block in K direction.
K2	Coordinate of last region block in K direction.
DZ	New value of *z* direction gross thickness for region (ft).

NOTE: NUMDZ records must be read.

6. **I1, I2, J1, J2, K1, K2, DZNET**

Omit this record if NUMDZN = 0.

I1	Coordinate of first region block in I direction.
I2	Coordinate of last region block in I direction.
J1	Coordinate of first region block in J direction.
J2	Coordinate of last region block in J direction.
K1	Coordinate of first region block in K direction.
K2	Coordinate of last region block in K direction.
DZNET	New value of *z* direction net thickness for region (ft).

NOTE: NUMDZN records must be read.

19.1.3 Depths to Top of Grid Blocks

The coordinate system used in BOAST4D is defined so that values in the z (vertical) direction increase as the layer gets deeper. Thus, depths must be read as depths below the user-selected reference datum. Negative values will be read as heights above the datum.

1. **Heading**   Up to 80 characters.

2. **KEL**

   KEL   Control code for input of depth values.

KEL	Meaning
0	A single constant value is read for the depth to the top of all grid blocks in layer 1 (horizontal plane). Each layer is contiguous in this option. Depths to the top of grid blocks in layers below layer 1 are calculated by adding the layer thickness to the preceding layer top; thus Top (I, J, K + 1) = Top (I, J, K) + DZ (I, J, K)
1	A separate depth value must be read for each grid block in layer 1. Read II × JJ values. Each layer is contiguous in this option. Depths to the top of grid blocks in layers below layer 1 are calculated by adding the layer thickness to the preceding layer top; thus Top (I, J, K + 1) = Top (I, J, K) + DZ (I, J, K)
2	A separate depth value is read for each layer. Read KK values. Each layer is horizontal (layer cake) in this option.
3	A separate depth value is read for each grid block. Read II × JJ × KK values.

3. **ELEV**

   ELEV   Depth to top of grid block (ft).

   If KEL = 0, read one constant value.

   If KEL = 1, read II × JJ values (one for each block in layer 1).

   If KEL = 2, read KK values (one for each layer).

   If KEL = 3, read II × JJ × KK values (one for each block).

19.2 Seismic Velocity Parameters

19.2.1 Moduli and Grain Densities

1. **Heading**   Up to 80 characters.

2. **KKM, KKG, KMU, KRHO**

    KKM   Control code for input of the frame bulk modulus
               (evacuated porous rock).

    KKG   Control code for input of the grain bulk modulus (solid
               matrix material).

    KMU   Control code for input of the shear modulus (evacuated
               porous rock).

    KRHO   Control code for input of the grain density (solid matrix
               material).

Code	Value	Meaning
KKM	-1	Frame bulk moduli are the same for all blocks in the grid. Read only one value.
	0	A constant value of frame bulk modulus is read for each layer in the grid; each layer may have a different, but constant value. Read KK values.
	1	Frame bulk moduli are read for each block in the grid. Read II × JJ × KK values.
KKG	-1	Grain bulk moduli are the same for all blocks in the grid. Read only one value.
	0	A constant value of grain bulk modulus is read for each layer in the grid; each layer may have a different, but constant value. Read KK values.
	1	Grain bulk moduli are read for each block in the grid. Read II × JJ × KK values.

Code	Value	Meaning
KMU	-1	Shear moduli are the same for all blocks in the grid. Read only one value.
	0	A constant value of shear modulus is read for each layer in the grid; each layer may have a different, but constant value. Read KK values.
	1	Shear moduli are read for each block in the grid. Read II × JJ × KK values.
KRHO	-1	Grain densities are the same for all blocks in the grid. Read only one value.
	0	A constant value of grain density is read for each layer in the grid; each layer may have a different, but constant value. Read KK values.
	1	Grain densities are read for each block in the grid. Read II × JJ × KK values.

3. **KM**

    KM   Frame bulk modulus (psia).

    If KKM = -1, read one constant value.

    If KKM = 0, read KK values (one for each layer).

    If KKM = +1, read II × JJ × KK values (one for each block).

    NOTE: In the absence of relevant data, a value of 3×10^6 psia is a reasonable estimate.

4. **KG**

    KG   Grain bulk modulus (psia).

    If KKG = -1, read one constant value.

    If KKG = 0, read JJ values (one for each layer).

    If KKG = +1, read II × JJ values (one for each block).

    NOTE: In the absence of relevant data, a value of 3×10^6 psia is a reasonable estimate.

5. **MU**

    MU   Shear modulus (psia).

    If KMU = -1, read one constant value.

    If KMU = 0, read KK values (one for each layer).

    If KMU = +1, read II × JJ × KK values (one for each block).

    NOTE: In the absence of relevant data, a value of 3×10^6 psia is a reasonable estimate.

6. **RHOMA**

    RHOMA   Grain density (lbf/ft^3).

    If KRHO = -1, read one constant value.

    If KRHO = 0, read KK values (one for each layer).

    If KRHO = +1, read II × JJ × KK values (one for each block).

    NOTE: In the absence of relevant data, a value of 168 lbf/ft^3 (corresponding to 2.7 g/cm^3) is a reasonable estimate.

19.2.2 Modifications to Moduli and Grain Densities

1. **Heading**   Up to 80 characters.

2. **NUMKM, NUMKG, NUMMU, NUMRHO, IDCODE**

    NUMKM   Number of regions where frame bulk modulus (KM) is changed.

    NUMKG   Number of regions where grain bulk modulus (KG) is changed.

    NUMMU   Number of regions where shear modulus (MU) is changed.

    NUMRHO   Number of regions where grain density (RHO) is changed.

    IDCODE   = 0 means do not print the modified distributions;
             = 1 means print the modified distributions.

3. **I1, I2, J1, J2, K1, K2, KM**

    Omit this record if NUMKM = 0.

I1	Coordinate of first region block in I direction.
I2	Coordinate of last region block in I direction.
J1	Coordinate of first region block in J direction.
J2	Coordinate of last region block in J direction.
K1	Coordinate of first region block in K direction.
K2	Coordinate of last region block in K direction.
KM	New value of frame bulk modulus (psia).

    NOTE: NUMKM records must be read.

4. **I1, I2, J1, J2, K1, K2, KG**

    Omit this record if NUMKG = 0.

I1	Coordinate of first region block in I direction.
I2	Coordinate of last region block in I direction.
J1	Coordinate of first region block in J direction.
J2	Coordinate of last region block in J direction.
K1	Coordinate of first region block in K direction.
K2	Coordinate of last region block in K direction.
KG	New value of grain bulk modulus (psia).

    NOTE: NUMKG records must be read.

5. **I1, I2, J1, J2, K1, K2, MU**

    Omit this record if NUMMU = 0.

I1	Coordinate of first region block in I direction.
I2	Coordinate of last region block in I direction.
J1	Coordinate of first region block in J direction.
J2	Coordinate of last region block in J direction.
K1	Coordinate of first region block in K direction.
K2	Coordinate of last region block in K direction.
MU	New value of shear modulus.

    NOTE: NUMMU records must be read.

6. **I1, I2, J1, J2, K1, K2, RHO**

   Omit this record if NUMRHO = 0.

   I1   Coordinate of first region block in I direction.

   I2   Coordinate of last region block in I direction.

   J1   Coordinate of first region block in J direction.

   J2   Coordinate of last region block in J direction.

   K1   Coordinate of first region block in K direction.

   K2   Coordinate of last region block in K direction.

   RHO   New value of grain density (lbf/ft^3).

   NOTE: NUMRHO records must be read.

19.3 Porosity, Permeability, and Transmissibility Distributions

---

19.3.1 Porosity and Permeability

1. **Heading**   Up to 80 characters.

2. **KPH, KKX, KKY, KKZ**

   KPH   Control code for input of porosity.

   KKX   Control code for input of x direction permeability.

   KKY   Control code for input of y direction permeability.

   KKZ   Control code for input of z direction permeability.

Code	Value	Meaning
KPH	-1	The porosity is constant for all grid blocks. Read only one value.
	0	A constant value is read for each layer in the grid. Read KK values.
	1	A value is read for each block in the grid. Read II × JJ × KK values.
KKX	-1	The x direction permeability is constant for all grid blocks. Read only one value.
	0	A constant value is read for each layer in the grid. Read KK values.
	1	A value is read for each block in the grid. Read II × JJ × KK values.
KKY	-1	The y direction permeability is constant for all grid blocks. Read only one value.
	0	A constant value is read for each layer in the grid. Read KK values.
	1	A value is read for each block in the grid. Read II × JJ × KK values.
KKZ	-1	The z direction permeability is constant for all grid blocks. Read only one value.
	0	A constant value is read for each layer in the grid. Read KK values.
	1	A value is read for each block in the grid. Read II × JJ × KK values.

3.   **PHI**

PHI   Porosity (fraction).

If KPH =   -1, read one constant value.

If KPH =   0, read KK values (one for each layer).

If KPH =   +1, read II × JJ × KK values (one for each block).

4. **PERMX**

    PERMX   Permeability in x direction (md).

    If KKX =   -1, read one constant value.

    If KKX =   0, read KK values (one for each layer).

    If KKX =   +1, read II × JJ × KK values (one for each block).

5. **PERMY**

    PERMY   Permeability in y direction (md).

    If KKY =   -1, read one constant value.

    If KKY =   0, read KK values (one for each layer).

    If KKY =   +1, read II × JJ × KK values (one for each block).

6. **PERMZ**

    PERMZ   Permeability in z direction (md).

    If KKZ =   -1, read one constant value.

    If KKZ =   0, read KK values (one for each layer).

    If KKZ =   +1, read II × JJ × KK values (one for each block).

19.3.2 Modifications to Porosities and Permeabilities

1. **Heading**   Up to 80 characters.

2. **NUMP, NUMKX, NUMKY, NUMKZ, IPCODE**

    NUMP   Number of regions where porosity (PHI) is changed.

    NUMKX   Number of regions where x direction permeability (PERMX) is changed.

    NUMKY   Number of regions where y direction permeability (PERMY) is changed.

    NUMKZ   Number of regions where z direction permeability (PERMZ) is changed.

    IPCODE   = 0 means do not print the modified distributions; = 1 means print the modified distributions.

3. **I1, I2, J1, J2, K1, K2, VALPHI**

    Omit this record if NUMP = 0.

    I1   Coordinate of first region block in I direction.

    I2   Coordinate of last region block in I direction.

    J1   Coordinate of first region block in J direction.

    J2   Coordinate of last region block in J direction.

    K1   Coordinate of first region block in K direction.

    K2   Coordinate of last region block in K direction.

Code	Value	Meaning
NUMP	< 0	New value of porosity for region (fr).
	> 0	Multiply value of porosity by VALPHI.

NOTE: |NUMP| records must be read.

4. **I1, I2, J1, J2, K1, K2, VALKX**

    Omit this record if NUMKX = 0.

    I1   Coordinate of first region block in I direction.

    I2   Coordinate of last region block in I direction.

    J1   Coordinate of first region block in J direction.

    J2   Coordinate of last region block in J direction.

    K1   Coordinate of first region block in K direction.

    K2   Coordinate of last region block in K direction.

Code	Value	Meaning
NUMKX	< 0	New value of x direction permeability for region (md).
	> 0	Multiply value of x direction permeability by VALKX.

NOTE: |NUMKX| records must be read.

5. **I1, I2, J1, J2, K1, K2, VALKY**

   Omit this record if NUMKY = 0.

I1	Coordinate of first region block in I direction.
I2	Coordinate of last region block in I direction.
J1	Coordinate of first region block in J direction.
J2	Coordinate of last region block in J direction.
K1	Coordinate of first region block in K direction.
K2	Coordinate of last region block in K direction.

Code	Value	Meaning
NUMKY	< 0	New value of y direction permeability for region (md).
	> 0	Multiply value of y direction permeability by VALKY.

   NOTE: |NUMKY| records must be read.

6. **I1, I2, J1, J2, K1, K2, VALKZ**

   Omit this record if NUMKZ = 0.

I1	Coordinate of first region block in I direction.
I2	Coordinate of last region block in I direction.
J1	Coordinate of first region block in J direction.
J2	Coordinate of last region block in J direction.
K1	Coordinate of first region block in K direction.
K2	Coordinate of last region block in K direction.

Code	Value	Meaning
NUMKZ	< 0	New value of z direction permeability for region (md).
	> 0	Multiply value of z direction permeability by VALKZ.

   NOTE: |NUMKZ| records must be read.

19.3.3 Modifications to Transmissibilities

It is important to keep in mind the directional convention used in specifying transmissibility modifications. For example, in grid block (I, J, K):

TX(I, J, K) refers to flow across the boundary between blocks I-1 and I,
TY(I, J, K) refers to flow across the boundary between blocks J-1 and J, and
TZ(I, J, K) refers to flow across the boundary between blocks K-1 and K.

1. **Heading**   Up to 80 characters.

2. **NUMTX, NUMTY, NUMTZ, ITCODE**

     NUMTX   Number of regions where *x* direction transmissibility (TX) is changed.

     NUMTY   Number of regions where *y* direction transmissibility (TY) is changed.

     NUMTZ   Number of regions where *z* direction transmissibility (TZ) is changed.

     ITCODE   = 0 means do not print the modified distributions;
                       = 1 means print the modified distributions.

3. **I1, I2, J1, J2, K1, K2, VALTX**

Omit this record if NUMTX = 0.

     I1   Coordinate of first region block in I direction.

     I2   Coordinate of last region block in I direction.

     J1   Coordinate of first region block in J direction.

     J2   Coordinate of last region block in J direction.

     K1   Coordinate of first region block in K direction.

     K2   Coordinate of last region block in K direction.

     VALTX   Multiplier of *x* direction transmissibility for region.
                  NOTE: NUMTX records must be read.

4.   **I1, I2, J1, J2, K1, K2, VALTY**

     Omit this record if NUMTY = 0.

I1	Coordinate of first region block in I direction.
I2	Coordinate of last region block in I direction.
J1	Coordinate of first region block in J direction.
J2	Coordinate of last region block in J direction.
K1	Coordinate of first region block in K direction.
K2	Coordinate of last region block in K direction.
VALTY	Multiplier of y direction transmissibility for region.

     NOTE: NUMTY records must be read.

5.   **I1, I2, J1, J2, K1, K2, VALTZ**

     Omit this record if NUMTZ = 0.

I1	Coordinate of first region block in I direction.
I2	Coordinate of last region block in I direction.
J1	Coordinate of first region block in J direction.
J2	Coordinate of last region block in J direction.
K1	Coordinate of first region block in K direction.
K2	Coordinate of last region block in K direction.
VALTZ	Multiplier of z direction transmissibility for region.

     NOTE: NUMTZ records must be read.

19.4 Rock and PVT Regions

---

1.   **Heading**   Up to 80 characters.

2.   **NROCK, NPVT**

NROCK	Number of distinct Rock regions. A separate set of saturation dependent data must be entered for each Rock region.
NPVT	Number of distinct PVT regions. A separate set of pressure dependent data must be entered for each PVT region.

3. **Heading**   Up to 80 characters.
   Omit this record if NROCK = 1.

4. **NUMROK**
   Omit this record if NROCK = 1.
   NUMROK = 0   Enter Rock region value for each block.
   NUMROK > 0   Number of regions where the Rock region default value of 1 is changed.

5. **IVAL**
   Omit this record if NROCK = 1 or NUMROK > 0.
   IVAL   Array of Rock region values. Read II × JJ × KK values.

6. **I1, I2, J1, J2, K1, K2, IVAL**
   Omit this record if NROCK = 1 or NUMROK = 0.
   I1   Coordinate of first region block in I direction.
   I2   Coordinate of last region block in I direction.
   J1   Coordinate of first region block in J direction.
   J2   Coordinate of last region block in J direction.
   K1   Coordinate of first region block in K direction.
   K2   Coordinate of last region block in K direction.
   IVAL   Number of the saturation dependent data set to be assigned to this Rock region and IVAL ≤ NROCK.
   NOTE: NUMROK records must be read.

7. **Heading**   Up to 80 characters.
   Omit this record if NPVT = 1.

8. **NUMPVT**
   Omit this record if NPVT = 1.
   NUMPVT = 0   Enter PVT region value for each block.
   NUMPVT > 0   Number of regions where the PVT region default value of 1 is changed.

9. **IVAL**

    Omit this record if NPVT = 1 or NUMPVT > 0.

    IVAL   Array of PVT region values. Read II × JJ × KK values.

10. **I1, I2, J1, J2, K1, K2, IVAL**

    Omit this record if NPVT = 1 or NUMPVT = 0.

    I1   Coordinate of first region block in I direction.

    I2   Coordinate of last region block in I direction.

    J1   Coordinate of first region block in J direction.

    J2   Coordinate of last region block in J direction.

    K1   Coordinate of first region block in K direction.

    K2   Coordinate of last region block in K direction.

    IVAL   Number of the pressure dependent data set to be assigned
                  to this PVT region and IVAL ≤ NPVT.

                  NOTE: NUMPVT records must be read.

19.5 Relative Permeability and Capillary Pressure Tables

The following saturation dependent data should be entered a total of NROCK times – one set of records for each defined Rock region.

1. **Heading**   Up to 80 characters.

2.   **SAT1   KROW1   KRW1   KRG1   KROG1   PCOW1   PCGO1**
         ⋮
       SATn   KROWn   KRWn   KRGn   KROGn   PCOWn   PCGOn

SAT   Phase saturation (fr). Set SAT1 = 0.0 and SATn = 1.0.

KROW   Oil relative permeability for oil-water system (fr).

KRW   Water relative permeability for oil-water system (fr).

KRG   Gas relative permeability for gas-oil system (fr).

KROG   Oil relative permeability for gas-oil system (fr).

PCOW   Oil/water capillary pressure (psi).

PCGO   Gas/oil capillary pressure (psi).

            NOTE: SAT refers to the saturation of each particular phase. For example, in a data line following SAT = 0.2 we have

KROW   Oil relative permeability at 20% oil saturation.

KRW   Water relative permeability at 20% water saturation.

KRG   Gas relative permeability at 20% gas saturation.

KROG   Oil relative permeability at 20% liquid (irreducible water plus oil) saturation.

PCOW   Oil/water capillary pressure at 20% water saturation.

PCGO   Gas/oil capillary pressure at 20% gas saturation.

            NOTE: KROG is used only when a three-phase oil relative permeability is calculated (ITHREE = 1 in Record 4 below). Capillary pressures are defined as PCOW = Po - Pw and PCGO = Pg - Po where Po, Pw, and Pg are the oil, water and gas phase pressures respectively.

3.   **Heading**   Up to 80 characters.

4.   **ITHREE, SWR**

ITHREE   Code specifying desired relative permeability option.

SWR   Irreducible water saturation (fraction).

Code	Value	Meaning
ITHREE	0	Oil relative permeability read from the relative permeability data for the two phase water/oil system.
	1	Oil relative permeability calculated from Stone's three-phase relative permeability model

Repeat records 1 to 4 a total of NROCK times.

19.6 Fluid PVT Tables

The following pressure dependent data should be entered a total of NPVT times – one set of records for each defined PVT region.

1.   **Heading**   Up to 80 characters.

2.   **PBO, PBODAT, PBGRAD**

PBO   Initial bubble point pressure (psia).

PBODAT   Depth at which PBO applies (ft).

PBGRAD   Constant bubble point pressure gradient (psia/ft).

3.   **Heading**   Up to 80 characters.

4.   **VSLOPE, BSLOPE, RSLOPE, PMAX, IREPRS**

VSLOPE   Slope of the oil viscosity versus pressure curve for undersaturated oil, i.e. for pressures above PBO. The slope ($\Delta\mu_o/\Delta P_o$) should be in cp/psia.

BSLOPE   Slope of the oil formation volume factor versus pressure curve for undersaturated oil. The slope ($\Delta B_o/\Delta P_o$) should be in RB/STB/psia and should be negative or zero. BSLOPE is not the same as the undersaturated oil compressibility.

RSLOPE   Slope of the solution gas-oil ratio versus pressure curve. The slope ($\Delta R_{so}/\Delta P_o$) should be in SCF/STB/psia and is normally zero.

PMAX   Maximum pressure entry for all PVT tables (psia).

IREPRS   = 0; constant bubble point pressure.
= 1; estimate variable bubble point pressure.

5.   **Heading**   Up to 80 characters; oil table follows.

6.   **P1**   **MUO1**   **BO1**   **RSO1**
      ⋮

PMAX   **MUO(PMAX)**   **BO(PMAX)**   **RSO(PMAX)**

P   Pressure (psia). Pressures must be in ascending order from P1 (normally 14.7 psia) to PMAX. The last table entry must be PMAX.

MUO   Saturated oil viscosity (cp).

BO   Saturated oil formation volume factor (RB/STB).

RSO   Saturated oil solution gas-oil ratio (SCF/STB).
NOTE: Oil properties must be entered as saturated oil over the entire pressure range.

7.   **Heading**   Up to 80 characters; water table follows.

8. **P1**   **MUW1**   **BW1**   **RSW1**
   ⋮

   PMAX   **MUW(PMAX)**   **BW(PMAX)**   **RSW(PMAX)**

   P   Pressure (psia). Pressures must be in ascending order from P1 (normally 14.7 psia) to PMAX. The last table entry must be PMAX.

   MUW   Water viscosity (cp).

   BW   Water formation volume factor (RB/STB).

   RSW   Water solution gas-water ratio (SCF/STB).

   NOTE: It is usually assumed in black oil simulations that the solubility of gas in water can be neglected. In this case, set RSW = 0.0 for all pressures.

9. **Heading**   Up to 80 characters.

10. **KGCOR**

Code	Value	Meaning
KGCOR	0	Read gas and rock properties table
	1	Activate gas correlation option and read rock compressibility vs pressure table

11. **Heading**   Up to 80 characters; gas table follows.

12. **P1   MUG1   BG1   PSI1   CR1**

⋮

PMAX   MUG(PMAX) BG(PMAX)   PSI(PMAX) CR(PMAX)
Omit this record if KGCOR = 1

P   Pressure (psia). Pressures must be in ascending order from
            P1 (normally 14.7 psia) to PMAX. The last table entry
            must be PMAX.

MUG   Gas viscosity (cp).

BG   Gas formation volume factor (RCF/SCF).

PSI   Gas pseudo-pressure ($psia^2/cp$).

CR   Rock compressibility (1/psia).

13. **KODEA, MPGT, TEM, SPG**
Omit this record if KGCOR = 0.

KODEA   Gas composition option (see Chapter 25.3).

MPGT   Number of gas PVT table entries ($1 < MPGT \leq 25$).

TEM   Reservoir temperature (°F).

SPG   Gas specific gravity (air = 1.0).

14. **FRCI**
Omit this record if KGCOR = 0.

FRCI   Component mole fraction of gas. Read 12 entries in the
            following order.

FRCI(I)	Component I	FRCI(I)	Component I
1	H_2S	7	iC_4
2	CO_2	8	nC_4
3	N_2	9	iC_5
4	C_1	10	nC_5
5	C_2	11	C_6
6	C_3	12	C_{7+}

15. **PRSCI, TEMCI, RMWTI**

    Omit this record if KGCOR = 0 or if KODEA ≠ 4.

    PRSCI   Critical pressure (psia).

    TEMCI   Critical temperature (°R).

    RMWTI   Molecular weight.

16. **Heading**   Up to 80 characters; rock compressibility table follows.
    Omit this record if KGCOR = 0.

17. **P1**   **CR1**

       ⋮

    PMAX   **CR(PMAX)**

    Omit this record if KGCOR = 0.

Option	Code	Meaning
Constant rock compressibility	PMAX	Maximum table pressure (psia) from record 4.
NOTE: Enter 1 record.	CR	Rock compressibility (1/psia)
Pressure dependent rock compressibility	P	Pressure (psia). Pressures must be in ascending order from P1 (normally 14.7 psia) to PMAX. The last table entry must be PMAX.
NOTE: Enter MPGT records.	CR	Rock compressibility (1/psia)

18. **Heading**   Up to 80 characters.

19. **RHOSCO, RHOSCW, RHOSCG**

    RHOSCO   Stock tank oil density (lb/cu ft).

    RHOSCW   Stock tank water density (lb/cu ft).

    RHOSCG   Gas density at standard conditions (lb/cu ft).

             NOTE: At standard conditions (14.7 psia and 60 degrees F for oilfield units) pure water has a density of 62.4 lb/cu ft and air has a density of 0.0765 lb/cu ft.

Repeat records 1 through 19 a total of NPVT times.

19.7 Pressure and Saturation Initialization

1.   **Heading**   Up to 80 characters.

2.   **KPI, KSI, PDATUM, GRAD**

    KPI   Pressure initialization code.

    KSI   Saturation initialization code.

    PDATUM   Depth to pressure datum (ft).

    GRAD   Estimated pressure gradient (psia/ft) for pressure corrections to PDATUM. If GRAD = 0, a map of pressures corrected to PDATUM will not be printed. If GRAD ≠ 0, a map of pressures corrected to PDATUM will be printed using pressure gradient GRAD.

Code	Value	Meaning
KPI	0	Read II × JJ × KK pressures (one for each block).
	1	Equilibrium pressure initialization. Requires pressures and depths at the OWC and GOC.
KSI	0	Read II × JJ × KK oil saturations (one for each block) and II × JJ × KK water saturations. Gas saturations will be calculated by the program.
	1	Gravity segregated oil, water and gas saturation initialization.

    NOTE: Options KPI and KSI may be used to prepare a restart data file.

3. **PO**

   Omit this record if KPI = 1.

   PO   Oil phase pressure (psia). Read II × JJ × KK values.

4. **PWOC, WOC, PGOC, GOC**

   Omit this record if KPI = 0.

   PWOC   Pressure at the water-oil contact (psia).

   WOC   Depth to the water-oil contact (ft below datum).

   PGOC   Pressure at the gas-oil contact (psia).

   GOC   Depth to the gas-oil contact (ft below datum).

                 NOTE: Repeat this record a total of NROCK times – one record for each Rock region.

5. **SO**

   Omit this record if KSI = 1.

   SO   Oil saturation array (fraction). Read II × JJ × KK values.

6. **SW**

   Omit this record if KSI = 1.

   SW   Water saturation array (fraction). Read II × JJ × KK values.

7. **SOI, SGI, SOR**

   Omit this record if KSI = 0.

   SOI   Initial oil saturation for the oil-water zone to be assigned to all blocks in the rock region (fraction). Initial water saturation in the oil-water zone is 1 - SOI.

   SGI   Initial gas saturation for the gas-water zone to be assigned to all blocks in the rock region (fraction). Initial water saturation in the gas-water zone is 1 - SGI.

   SOR   Irreducible oil saturation to be assigned to all blocks in the rock region (fraction). If SOR > 0, calculated So will be set to 0 when So < SOR. Water and gas saturations are then renormalized.

NOTE: Repeat this record a total of NROCK times – one record for each Rock region.

19.8 Run Control Parameters

---

1. **Heading**   Up to 80 characters.

2. **NMAX, FACT1, FACT2, TMAX, WORMAX, GORMAX, PAMIN, PAMAX**

    NMAX   Maximum number of time steps allowed.

    FACT1   Factor for increasing time step size using automatic time step control. FACT1 = 1.0 for fixed time step size. A common value for FACT1 is 1.25.

    FACT2   Factor for decreasing time step size using automatic time step control. FACT2 = 1.0 for fixed time step size. A common value for FACT2 is 0.5.

    TMAX   Maximum elapsed time to be simulated (days); the run will be terminated when the time exceeds TMAX.

    WORMAX Maximum allowed water-oil ratio for a producing oil well (STB/STB); WORMAX ≥ 0.

    GORMAX Maximum allowed gas-oil ratio for a producing oil well (SCF/STB); GORMAX ≥ 0.

    PAMIN   Minimum field average pressure (psia); the run will be terminated when the pore volume weighted average reservoir pressure < PAMIN.

    PAMAX   Maximum field average pressure (psia); the run will be terminated when the pore volume weighted average reservoir pressure > PAMAX.

    NOTE: PAMIN and PAMAX should be within the range of pressures covered by the fluid PVT tables discussed in Chapter 19.6.

3. **WOROCK**

   Omit this record if WORMAX ≠ 0.

   WOROCK Maximum WOR allowed in the corresponding Rock region.

   NOTE: If a well is completed in more than one Rock region, the largest maximum WOR which applies to the Rock regions penetrated by the well will be used as the WOR control for that well. Enter NROCK records – one for each Rock region.

4. **GOROCK**

   Omit this record if GORMAX ≠ 0.

   GOROCK Maximum GOR allowed in the corresponding Rock region.

   NOTE: If a well is completed in more than one Rock region, the largest maximum GOR which applies to the Rock regions penetrated by the well will be used as the GOR control for that well. Enter NROCK records – one for each Rock region.

19.9 Solution Method Specification

1. **Heading** Up to 80 characters.

2. **KSOL, MITR, OMEGA, TOL, TOL1, DSMAX, DPMAX, NUMDIS**

    KSOL   Solution method code.

    MITR   Maximum number of LSOR iterations per time step. A typical value is 100.

    OMEGA   Initial LSOR acceleration parameter. Values of OMEGA should be between 1.0 and 2.0. A typical initial value is 1.5.

    TOL   Maximum acceptable pressure change for convergence of LSOR iterations (psia). A typical value is 0.1.

    TOL1   Parameter for determining when to change OMEGA. A typical value is 0.001. If TOL1 = 0.0, the initial value of OMEGA will be used for the entire run.

    DSMAX   Maximum saturation change allowed per time step (fraction). The time step size will be reduced by FACT2 if the saturation change of a phase in any grid block exceeds DSMAX during a time step. A typical value for DSMAX is 0.05.

    DPMAX   Maximum pressure change allowed per time step (psia). The time step size will be reduced by FACT2 if the pressure change in any grid block exceeds DPMAX during a time step. A typical value of DPMAX is 100 psia.

    NUMDIS   Code for controlling numerical dispersion

Code	Value	Meaning
	1	1D Tridiagonal Algorithm. Use with 1D problems and 0D (tank) problems, i.e. when II = JJ = KK = 1.
	2	Direct solution band algorithm. Use with 2D and 3D problems.
KSOL	3	LSORX - Iterative matrix solver with direct solver in x direction.
	4	LSORY - Iterative matrix solver with direct solver in y direction.
	5	LSORZ - Iterative matrix solver with direct solver in z direction.

Code	Value	Meaning
NUMDIS	1	Single-point upstream weighting.
	2	Two-point upstream weighting.

19.10 Analytic Aquifer Models

1. **Heading**   Up to 80 characters.

2. **IAQOPT**
   IAQOPT   Analytic aquifer model code.

Code	Value	Meaning
IAQOPT	0	No analytic aquifer model
	1	Pot aquifer model (small and bounded aquifer)
	2	Steady-state aquifer model (constant aquifer pressure)
	3	Carter-Tracy aquifer model: Re/Rw = 1.5
	4	Carter-Tracy aquifer model: Re/Rw = 2.0
	5	Carter-Tracy aquifer model: Re/Rw = 3.0
	6	Carter-Tracy aquifer model: Re/Rw = 4.0
	7	Carter-Tracy aquifer model: Re/Rw = 5.0
	8	Carter-Tracy aquifer model: Re/Rw = 6.0
	9	Carter-Tracy aquifer model: Re/Rw = 8.0
	10	Carter-Tracy aquifer model: Re/Rw = 10.0
	11	Carter-Tracy aquifer model: Re/Rw = ∞

NOTE: Only one aquifer model option (IAQOPT) may be selected for a given run. Different aquifer influx strengths may be specified for a given aquifer.

3.   **NAQEN**

Omit this record if IAQOPT ≠ 1.

NAQEN   Number of regions containing a pot aquifer.

4.   **I1, I2, J1, J2, K1, K2, POT**

Omit this record if IAQOPT ≠ 1.

I1   Coordinate of first region block in I direction.

I2   Coordinate of last region block in I direction.

J1   Coordinate of first region block in J direction.

J2   Coordinate of last region block in J direction.

K1   Coordinate of first region block in K direction.

K2   Coordinate of last region block in K direction.

POT   Pot aquifer strength (SCF/psia).

         NOTE: NAQEN records must be read.

5.   **NAQEN**

Omit this record if IAQOPT ≠ 2.

NAQEN   Number of regions containing a steady-state aquifer.

6.   **I1, I2, J1, J2, K1, K2, SSAQ**

Omit this record if IAQOPT ≠ 2.

I1   Coordinate of first region block in I direction.

I2   Coordinate of last region block in I direction.

J1   Coordinate of first region block in J direction.

J2   Coordinate of last region block in J direction.

K1   Coordinate of first region block in K direction.

K2   Coordinate of last region block in K direction.

SSAQ   Steady-state aquifer strength (SCF/day/psia).

         NOTE: NAQEN records must be read.

7.   **NAQREG**

Omit this record if IAQOPT < 3.

NAQREG   Number of Carter-Tracy aquifer parameter regions.

8. **AQCR, AQCW, AQMUW, AQK, AQPHI, AQH, AQS, AQRE**
    Omit this record if IAQOPT < 3.

    AQCR   Aquifer rock compressibility (1/psia).

    AQCW   Aquifer water compressibility (1/psia).

    AQMUW   Aquifer water viscosity (cp).

    AQK   Aquifer permeability (md).

    AQPHI   Aquifer porosity (fraction).

    AQH   Aquifer net thickness (ft).

    AQS   Aquifer to reservoir boundary interface (fraction). A value of 0 implies there is no boundary (hence no influx); a value of 1 implies that the aquifer surrounds the grid block.

    AQRE   External aquifer radius (ft).

9. **NAQEN**
    Omit this record if IAQOPT < 3.

    NAQEN   Number of regions containing a Carter-Tracy aquifer.

10. **I1, I2, J1, J2, K1, K2**
    Omit this record if IAQOPT < 3.

    I1   Coordinate of first region block in I direction.

    I2   Coordinate of last region block in I direction.

    J1   Coordinate of first region block in J direction.

    J2   Coordinate of last region block in J direction.

    K1   Coordinate of first region block in K direction.

    K2   Coordinate of last region block in K direction.

    NOTE: NAQEN lines must be read. Repeat records 8 through 10 a total of NAQREG times.

Chapter 20

Recurrent Data

Recurrent data records are read periodically during the course of the simulation run. These data include the location and specification of wells in the model, changes in well completions and field operations over time, a schedule of well rate and/or pressure performance over time, time step control information for advancing the simulation through time, and controls on the type and frequency of printout information provided by the simulator.

1. **Major Heading** Up to 80 characters.

   NOTE: This record signifies the start of the recurrent data section.

20.1 Time Step and Output Control

Time step and output control records must be read to start the simulation.

1. **Heading** Up to 80 characters.

2. **IWLCNG, IOMETH**

   IWLCNG Controls reading of well information.

   IOMETH Controls program output and well scheduling.

Code	Value	Meaning
IWLCNG	0	Do not read well information
	1	Read well information
IOMETH	≥ 1	Number of elapsed time values to be read on record 3. The program will print results to output files at these elapsed times and allow you to change well characteristics after the last elapsed time entered during this recurrent data period.

3. **FTIO**

   FTIO   Array containing total elapsed times at which output will occur (days). Up to 50 monotonically increasing values may be entered. The first entry must be greater than 0 and greater than the last entry of any previously completed recurrent data periods.

   NOTE: When the elapsed time of a run equals an FTIO value, the well and basic summary reports will be printed. Maps will also be printed according to the instructions given in record 4.

4. **IPMAP, ISOMAP, ISWMAP, ISGMAP, IPBMAP, IAQMAP**

   IPMAP   Control code for printing pressure array.

   ISOMAP   Control code for printing oil saturation array.

   ISWMAP   Control code for printing water saturation array.

   ISGMAP   Control code for printing gas saturation array.

   IPBMAP   Control code for printing bubble point pressure array.

   IAQMAP   Control code for printing aquifer influx array.

Code Value	Meaning
0	Do not print the array
1	Print the array
2	Print the array and a digital contour plot

5.   **IVPMAP, IZMAP, IRCMAP, IVSMAP, IVRMAP**

      IVPMAP   Control code for printing seismic compressional velocity (Vp) array.

      IZMAP   Control code for printing seismic acoustic impedance array.

      IRCMAP   Control code for printing seismic reflection coefficient array.

      IVSMAP   Control code for printing seismic shear velocity (Vs) array.

      IVRMAP   Control code for printing seismic velocity ratio Vp/Vs array.

Code Value	Meaning
0	Do not print the array
1	Print the array
2	Print the array and a digital contour plot

6.   **DT, DTMIN, DTMAX**

      DT   Starting time step size (days). DT may vary between DTMIN and DTMAX when automatic time step control is invoked.

      DTMIN   Minimum time step size allowed (days). A typical value is 1 day.

      DTMAX   Maximum time step size allowed (days). A typical value is 30 days.

20.2 Well Information

Omit this section if IWLCNG = 0.

1.   **Heading**   Up to 80 characters.

2. **NWELLN, NWELLO**

   NWELLN   Number of new wells for which complete well information is entered.

   NWELLO   Number of previously defined wells for which new rates and/or rate controls are entered.

3. **Heading**   Up to 80 characters.
   Omit this record if NWELLN = 0.

4. **WELLID**
   Omit this record if NWELLN = 0.
   WELLID   Well name with up to five characters.

5. **IDWELL, KONECT**
   Omit this record if NWELLN = 0.
   IDWELL   Well identification number. Each well should have a unique IDWELL number. If two or more wells have the same IDWELL number, the characteristics of the last well entered will be used.

   KONECT   Total number of grid blocks connected to well IDWELL.

6. **I, J, K, PID, PWF**
   Omit this record if NWELLN = 0.
   I x coordinate of grid block containing well.
   J y coordinate of grid block containing well.
   K z coordinate of grid block containing well.
   PID   Layer flow index for grid block.
   PWF   Flowing bottom-hole pressure for block (psia). This value is used only if KIP is negative for this well.

              NOTE: KONECT records must be read. PID for a vertical well can be estimated as

$$PID = 0.00708 \ \frac{Kh}{\ln\left(\dfrac{r_o}{r_w}\right) + S}$$

where

$$r_o \approx 0.14 \, (DX^2 + DY^2)^{\frac{1}{2}}$$

and

K	$=$	layer absolute permeability (md)
h	$=$	layer thickness (ft)
DX	$=$	x direction grid block length (ft)
DY	$=$	y direction grid block length (ft)
r_w	$=$	wellbore radius (ft)
r_o	$=$	equivalent well block radius (ft)
S	$=$	layer skin factor

Deviated (slanted) and horizontal wells may be represented by calculating an appropriate PID and specifying grid block locations that model the expected well trajectory. For example, a horizontal well that is aligned in the x direction will have constant J and K indices, and index I will vary if there is more than one connection.

To shut in a connection, set that connection PID to 0. To shut in a well, set all of its connection PID values to zero.

7. **KIP, QO, QW, QG, QT**
   Omit this record if NWELLN = 0.
   KIP   Code for specifying well operating characteristics.
   Rate Controlled Well (KIP > 0):
       QO   Oil rate (STB/D).
       QW Water rate (STB/D).
       QG   Gas rate (MSCF/D).
       QT   Total fluid voidage rate (RB/D).
             NOTE: The total fluid rate given by QT is the oil plus water plus gas production for the well or the total reservoir voidage rate at reservoir conditions. For multi-layer systems, QT is a target rate.

BHP Controlled Production Well with Optional Rate Constraints
(KIP = -1):

QO Minimum oil production rate required (STB/D).

QW Maximum oil production rate allowed (STB/D).

QG 0.0

QT Maximum liquid withdrawal rate allowed (STB/D).

NOTE: Rate constraints are not activated if the corresponding rate equals zero.

BHP Controlled Water Injection Well with Optional Rate Constraints
(KIP = -2):

QO 0.0

QW Maximum water injection rate allowed (STB/D).

QG 0.0

QT 0.0

NOTE: QW should be a negative number or zero. The rate constraint is not activated if QW = 0.

BHP Controlled Gas Injection Well with Optional Rate Constraints
(KIP = -3):

QO 0.0

QW 0.0

QG Maximum gas injection rate allowed (MSCF/D).

QT 0.0

NOTE: QG should be a negative number or zero. The rate constraint is not activated if QG = 0.

Gas Production Well (KIP = -4):

QO 0.0

QW 0.0

QG 0.0

QT 0.0

NOTE: Sign conventions for rates:

Negative rates indicate fluid injection.

Positive rates indicate fluid production.

Summary of KIP Values	
Code	**Meaning**
3	Gas well - injection rate specified
2	Water well - injection rate specified
1	Production well - rate specified ♦ Oil rate specified: $QO > 0$, $QW = QG = QT = 0$ ♦ Water rate specified: $QW > 0$, $QO = QG = QT = 0$ ♦ Gas rate specified: $QG > 0$, $QO = QW = QT = 0$ ♦ Total rate specified: $QT > 0$, $QO = QW = QG = 0$
-1	Oil and/or water production well - PI and FBHP control
-2	Water well - PI and FBHP control
-3	Gas injection well - PI and FBHP control
-4	Gas production well - LIT representation

8.   **ALIT, BLIT**
Omit this record if NWELLN = 0 or KIP ≠ -4.
ALIT   "a" coefficient of LIT gas well analysis.
BLIT   "b" coefficient of LIT gas well analysis.

NOTE: Records 4 through 8 should be repeated NWELLN times.

9.   **Heading**   Up to 80 characters.
Omit this record if NWELLO = 0.

10.   **WELLID**
Omit this record if NWELLO = 0.
WELLID   Well name with up to five characters.

11. **IDWELL, KONECT**

    Omit this record if NWELLO = 0.

    IDWELL   Well identification number. Each well should have a unique IDWELL number. If two or more wells have the same IDWELL number, the characteristics of the last well entered will be used.

    KONECT   Total number of grid blocks connected to well IDWELL.

12. **I, J, K, PID, PWF**

    Omit this record if NWELLO = 0.

    I x coordinate of grid block containing well.

    J y coordinate of grid block containing well.

    K z coordinate of grid block containing well.

    PID   Layer flow index for grid block.

    PWF   Flowing bottom-hole pressure for block (psia). This value is used only if KIP is negative for this well.

    NOTE: KONECT records must be read.

13. **KIP, QO, QW, QG, QT**

    Omit this record if NWELLO = 0.

    KIP   Code for specifying well operating characteristics. See record 6 for a description of the KIP options.

14. **ALIT, BLIT**

    Omit this record if NWELLO = 0 or KIP ≠ -4.

    ALIT   "a" coefficient of LIT gas well analysis.

    BLIT   "b" coefficient of LIT gas well analysis.

NOTE: Records 10 through 14 should be repeated NWELLO times.

Chapter 21

Program Output Evaluation

You are given the option at the start of a BOAST4D run to direct output to either the screen or to a set of files. It is often worthwhile to send output to the screen when first building and debugging a data set. BOAST4D will abort at the point in the data set where it encounters improperly entered data. For evaluating run results, it is preferable to send output to files. In this case, a one line time step summary is sent to the screen each time step so that you can monitor the progress of a run. All output files are in ASCII format.

A run may be aborted by typing <Cntl> C. You may then choose to terminate the job.

21.1 Initialization Data

BOAST4D outputs the following initialization data in ASCII file BTEMP.OUT:

- Grid block sizes
- Node midpoint elevations
- Porosity distributions
- Permeability distributions
- Rock and PVT region distributions
- Relative permeability and capillary pressure tables
- PVT tables
- Slopes calculated from PVT data
- Time step control parameters
- Analytic aquifer model selection
- Initial fluid volumes-in-place

- ◆ Initial pressure and saturation arrays
- ◆ Initial seismic velocities array
- ◆ Initial acoustic impedance array
- ◆ Initial well information

Other output can be obtained at your request. For example, if a modification option is invoked, you may print out the altered array. It is worthwhile to do this as a check on the input changes.

21.2 Recurrent Data

All output files are ASCII files so that they may be read by a variety of commercially available spreadsheets. BOAST4D output may then be manipulated using spreadsheet options. This is especially useful for making plots or displaying array data. Different output files are defined so that simulator output file sizes are more manageable. The output files are designed to contain information that is logically connected, e.g. well data in one file, reservoir property distributions in another file. The different output files are described below.

21.2.1 Time Step Summary File – BTEMP.TSS

A one line time step summary is automatically printed out as a record of the progress of the run. This summary provides you with necessary information for evaluating the stability of the solution as a function of time. Significant oscillations in GOR or WOR, or large material balance errors are indicative of simulation problems and should be corrected. A smaller time step through the difficult period is often sufficient to correct IMPES instabilities.

21.2.2 Run Summary And Plot File – BTEMP.PLT

The run summary file contains a concise summary of total field production and injection, and fieldwide aquifer influx. The WOR and GOR are ratios of total producing fluid rates. Consequently these ratios are comparable to observed fieldwide ratios.

The output quantities include: cumulative production of oil, water and gas; cumulative injection of water and gas; pore volume weighted average pressure; aquifer influx rate and cumulative aquifer influx; and fieldwide WOR and GOR values. These quantities are output as functions of time and time step number.

21.2.3 Well Report File – BTEMP.WEL

Rates and cumulative production/injection data for each layer of each well are summarized in the well report at times you specify. Field totals are also included.

21.2.4 Distribution Arrays File – BTEMP.OUT

You may output the following arrays whenever desired: pressure, saturations, bubble point pressure, cumulative aquifer influx, compressional velocity, acoustic impedance and seismic reflection coefficient. Output arrays may be used as input pressure and saturation distributions for restarting a run.

It is usually unnecessary to print all of the arrays. To avoid excessive output and correspondingly large output files, you should be judicious in deciding which arrays are printed. In addition to arrays, you may wish to output digital contour plots.

Digital contour plots provide a simplified picture of the physical parameter distribution. The plot subroutine finds the minimum (AMIN) and maximum (AMAX) values of the array APLOT. A new array AOUT is constructed using the normalized parameter values given by

$$AV = (APLOT(I, J, K) - AMIN)/ADIF$$

where ADIF = AMAX - AMIN > 0.001. The values of AOUT are defined as follows:

AOUT	Meaning (±0.05)
–	AV < 0.05
1	AV = 0.10
2	AV = 0.20
3	AV = 0.30
4	AV = 0.40
5	AV = 0.50
6	AV = 0.60
7	AV = 0.70
8	AV = 0.80
9	AV = 0.90
T	AV > 0.95

Digital contour plots highlight changes in parameter values and let you visually monitor such items as saturation fronts, movements of pressure pulses, and changes in acoustic impedance. The output array AOUT is printed so that it can be used for drawing a rough contour plot.

Chapter 22

Example Input Data Sets

FILE	GRID II × JJ × KK	MODEL TYPE	REMARKS
EXAM1.DAT	1 × 1 × 1	Material Balance	Primary depletion of an under-saturated oil reservoir (high GOR)
EXAM2.DAT	1 × 1 × 4	1D Vertical	Primary depletion of an under-saturated oil reservoir (moderate GOR)
EXAM3.DAT	10 × 1 × 1	1D Horizontal	Buckley-Leverett waterflood
EXAM4.DAT	9 × 9 × 1	2D Areal	Primary depletion of an undersaturated oil reservoir (high GOR)
EXAM5.DAT	10 × 1 × 4	2D Cross- section	Multi-layer waterflood of an undersaturated oil reservoir (high GOR)
EXAM6.DAT	9 × 9 × 2	3D	5-spot waterflood of an under-saturated oil reservoir (high GOR)
EXAM7.DAT	10 × 10 × 3	3D	Gas injection into undersaturated oil reservoir (high GOR) - Odeh example
EXAM8.DAT	9 × 9 × 2	3D	Depletion of gas reservoir
EXAM9.DAT	9 × 9 × 2	3D	Depletion of gas reservoir with aquifer support
EXAM10.DAT	10 × 8 × 4	3D	Depletion of a faulted oil reservoir with multiple PVT and ROCK regions
EXAM11.DAT	10 × 1 × 2	2D Cross- section	Depletion of gas reservoir with aquifer support

EXAMPLE INPUT DATA SET

The following data set is presented to illustrate the BOAST4D input file format. Additional spacing has been provided between some lines to improve data set readability. The actual BOAST4D data set should contain no blank lines between records.

```
---------------------------- BEGINNING OF DATA SET ----------------------------------
 PRIMARY DEPLETION OF AN OIL RESERVOIR - VERTICAL COLUMN MODEL
GRID DIMENSIONS
  1,  1,  4
GRID BLOCK LENGTHS
 -1 -1  0  0
  2000.0
  1200.0
   2*50.0  2*60.0
   2*36.0  2*38.0
GRID BLOCK LENGTH MODIFICATIONS
  0,  0,  0, 0,  0
DEPTH TO TOP OF UPPER SAND
  2
  9330
  9380
  9430
  9490
MODULI AND ROCK DENSITY
  -1 -1 -1 -1
  3E6
  3E6
  3E6
  168
MODULI AND ROCK DENSITY MODIFICATIONS
  0 0 0 0 0
POROSITY AND PERMEABILITY DISTRIBUTIONS
   0   0   0   0
   2*0.20  2*0.25
   2*75  2*250
   2*75  2*250
   2*7.5  2*25
POROSITY AND PERMEABILITY MODIFICATION CARDS
  0,  0,  0,  0,  0
TRANSMISSIBILITY MOD. - NO FLOW BETWEEN LAYERS 2 AND 3
   0,  0,  1,  0
   1  1  1  1  3  3  0.0
ROCK AND PVT REGIONS
   1,  1
```

SAT	KRO	KRW	KRG	KROG	PCOW	PCGO
0.00	0.00	0.00	0.00	0.0	0.0	0.0
0.03	0.00	0.00	0.00	0.0	0.0	0.0
0.05	0.00	0.00	0.02	0.0	0.0	0.0
0.10	0.00	0.00	0.09	0.0	0.0	0.0
0.15	0.00	0.00	0.16	0.0	0.0	0.0
0.20	0.00	0.00	0.24	0.0	0.0	0.0
0.25	0.00	0.00	0.33	0.0	0.0	0.0
0.30	0.0001	0.00	0.43	0.0	0.0	0.0
0.35	0.001	0.005	0.55	0.0	0.0	0.0
0.40	0.01	0.010	0.67	0.0	0.0	0.0
0.45	0.03	0.017	0.81	0.0	0.0	0.0
0.50	0.08	0.023	1.00	0.0	0.0	0.0
0.55	0.18	0.034	1.00	0.0	0.0	0.0
0.60	0.32	0.045	1.00	0.0	0.0	0.0
0.65	0.59	0.064	1.00	0.0	0.0	0.0
0.70	1.00	0.083	1.00	0.0	0.0	0.0
0.80	1.00	0.12	1.00	0.0	0.0	0.0
0.90	1.00	0.12	1.00	0.0	0.0	0.0
1.00	1.00	0.12	1.00	0.0	0.0	0.0

ITHREE   SW(IRR.)
   0,   0.30
PBO   PBODAT   PBGRAD
2514.7, 9200.0,   0.0
VSLOPE   BSLOPE   RSLOPE   PMAX REPRS
0.000046,   -0.000023,   0.0,   6014.7, 0

OIL: P	MUO	BO	RSO
14.7	1.0400	1.0620	1.0
514.7,	0.9100,	1.1110,	89.0
1014.7,	0.8300,	1.1920,	208.0
1514.7,	0.7650,	1.2560,	309.0
2014.7,	0.6950,	1.3200,	392.0
2514.7,	0.6410,	1.3800,	457.0
3014.7,	0.5940,	1.4260,	521.0
4014.7,	0.5100,	1.4720,	586.0
5014.7,	0.4500,	1.4900,	622.0
6014.7,	0.4100,	1.5000,	650.0

```
WATER: P      MUW     BW        RSW
    14.7,     0.5000,  1.0190,  0.0
   514.7,     0.5005,  1.0175   0.0
  1014.7,     0.5010,  1.0160,  0.0
  1514.7,     0.5015,  1.0145,  0.0
  2014.7,     0.5020,  1.0130,  0.0
  2514.7,     0.5025,  1.0115,  0.0
  3014.7,     0.5030,  1.0100,  0.0
  4014.7,     0.5040,  1.0070,  0.0
  5014.7,     0.5050,  1.0040,  0.0
  6014.7,     0.5060,  1.0010,  0.0


GAS AND ROCK PROPERTIES
  0
P           MUG         BG         PSI    CR
    14.7,   0.008000,   0.935800,  0.0,   0.000003
   514.7,   0.011200,   0.035200,  0.0,   0.000003
  1014.7,   0.014000,   0.018000,  0.0,   0.000003
  1514.7,   0.016500,   0.012000,  0.0,   0.000003
  2014.7,   0.018900,   0.009100,  0.0,   0.000003
  2514.7,   0.020800,   0.007400,  0.0,   0.000003
  3014.7,   0.022800,   0.006300,  0.0,   0.000003
  4014.7,   0.026000,   0.004900,  0.0,   0.000003
  5014.7,   0.028500,   0.004000,  0.0,   0.000003
  6014.7,   0.030000,   0.003400,  0.0,   0.000003
RHOSCO     RHOSCW      RHOSCG
  46.244,   62.238,     0.0647
EQUILIBRIUM PRESSURE INIT. / CONSTANT SATURATION INIT.
    1,    1,   0,   0
  4000,  9600,  0,   8000
  0.70,  0,   0.25
NMAX    FACT1   FACT2   TMAX  WORMAX  GORMAX   PAMIN   PAMAX
1000,   1.50,   0.50,   365,   5.0,    500000,  1500,   6000
KSOL    MITR    OMEGA   TOL   TOL1    DSMAX   DPMAX   NUMDIS
1,      100,    1.50,   0.1,   0.001,  0.05,   100.0,   1
AQUIFER MODEL
  0
RECURRENT DATA
  *** DATA SET 1 - HISTORY ***
    1,   4
  91.25  182.5  273.75  365.0
  1,  1,  1,  0,  0,  1
  0,  0,  0,  0,  0
  5.0,   1.0,   10.0
WELL INFORMATION
  1    0
```

```
WELL P-1
P-1
  1   4
  1,  1,  1,  2.7   2600
  1,  1,  2,  2.7   2600
  1,  1,  3,  9.4   2600
  1,  1,  4,  9.4   2600
  1,   500.0,     0.0,     0.0,     0.0
```
------------------------------ END OF DATA SET ---------------------------------

Part IV

BOAST4D

Technical Supplement

Chapter 23

Simulator Formulation

BOAST4D is an implicit pressure-explicit saturation finite difference simulator. It can simulate isothermal Darcy flow in up to three dimensions. Reservoir fluids are described by up to three fluid phases (oil, gas, and water), whose physical properties are functions of pressure only. Solution gas may be present in both the oil and water phases.

23.1 Equations

The black oil simulator mass conservation equations for the oil, water and gas phases are derived in Chapter 32. They can be succinctly written as follows:

Oil

$$-\nabla \cdot \frac{\vec{v}_o}{B_o} - \frac{q_o}{\rho_{osc}} = \frac{\partial}{\partial t}\left(\phi \frac{S_o}{B_o}\right) \tag{23.1}$$

Water

$$-\nabla \cdot \frac{\vec{v}_w}{B_w} - \frac{q_w}{\rho_{wsc}} = \frac{\partial}{\partial t}\left(\phi \frac{S_w}{B_w}\right) \tag{23.2}$$

Gas

$$-\nabla \cdot \left[\frac{\vec{v}_g}{B_g} + \frac{R_{so}}{B_o} \vec{v}_o + \frac{R_{sw}}{B_w} \vec{v}_w \right] - \frac{q_g}{\rho_{gsc}}$$

$$= \frac{\partial}{\partial t} \left\{ \phi \left[\frac{S_g}{B_g} + \frac{R_{so}}{B_o} S_o + \frac{R_{sw}}{B_w} S_w \right] \right\}$$

(23.3)

Letting the subscript i denote o (oil), w (water), and g (gas), the symbols in Eqs. (23.1) to (23.3) are defined as follows:

B_i = formation volume factor of phase i

q_i = mass flow rate per unit reservoir volume of phase i

R_{so} = solubility of gas in oil

R_{sw} = solubility of gas in water

S_i = saturation of phase i

v_i = Darcy's velocity of phase i

ρ_{isc} = density of phase i at standard conditions

ϕ = porosity

Three additional equations – called auxiliary equations – are employed when solving the preceding fluid flow equations. They are

$$S_o + S_w + S_g = 1$$

(23.4)

$$P_{cow}(S_w) = P_o - P_w$$

(23.5)

$$P_{cgo}(S_g) = P_g - P_o$$

(23.6)

where P_i is the pressure of phase i, P_{cow} is the oil-water capillary pressure, and P_{cgo} is the gas-oil capillary pressure.

Darcy's velocity for phase i is

$$\vec{v}_i = -K \frac{k_{ri}}{\mu_i} \nabla \Phi_i$$

(23.7)

where K is a permeability tensor that is usually assumed to be diagonalized along its principal axes, k_{ri} and μ_i are relative permeability and viscosity of phase i respectively. The phase potentials Φ_i are given as functions of depth z by

$$\Phi_o = P_o - \frac{\rho_o z}{144}, \quad \Phi_w = P_o - P_{cow} - \frac{\rho_w z}{144}, \quad \Phi_g = P_o + P_{cgo} - \frac{\rho_g z}{144} \qquad (23.8)$$

Phase densities are calculated from input PVT data as

$$\rho_o = \frac{1}{B_o}[\rho_{osc} + R_{so}\rho_{gsc}], \quad \rho_w = \frac{1}{B_w}[\rho_{osc} + R_{sw}\rho_{gsc}], \quad \rho_g = \frac{\rho_{gsc}}{B_g} \qquad (23.9)$$

Expressions for rock and phase compressibilities are

$$c_r = \frac{1}{\phi}\frac{\partial \phi}{\partial P_o}, \quad c_g = -\frac{1}{B_g}\frac{\partial B_g}{\partial P_o},$$

$$c_o = -\left[\frac{1}{B_o}\frac{\partial B_o}{\partial P_o} - \frac{B_g}{B_o}\frac{\partial R_{so}}{\partial P_o}\right], \quad c_w = -\left[\frac{1}{B_w}\frac{\partial B_w}{\partial P_o} - \frac{B_g}{B_w}\frac{\partial R_{sw}}{\partial P_o}\right] \qquad (23.10)$$

These equations are discretized and solved numerically in BOAST4D. The procedure for solving these equations is outlined in Chapter 33.

23.2 Transmissibility

The simulator offers no-flow boundary conditions, which lets you stop flow between specified grid blocks in chosen directions. The no-flow conditions are implemented by setting transmissibilities at boundary interfaces to zero. The Transmissibility Modifications section in Chapter 19.3.3 describes the directional conventions for transmissibility in the model.

Flow between neighboring blocks is treated as a series application of Darcy's law. A transmissibility term is defined using the product of average values of relative permeability, absolute permeability, and cross-sectional area, divided by the product of viscosity and formation volume factor. The transmissibility to each phase is determined using a harmonic average calculation of the product of absolute permeability and cross-sectional area at the interface between neighboring blocks. An arithmetic average of phase viscosities and formation volume factors is used. The average relative permeability is determined using an upstream weighted averaging technique.

23.3 Coordinate Orientation

The BOAST4D reservoir model assumes a block-centered grid with the axes aligned using the right-handed coordinate reference illustrated in Figure 23-1. The top layer (K = 1) is shown. The second layer (K = 2) is below the K = 1 layer, and so on.

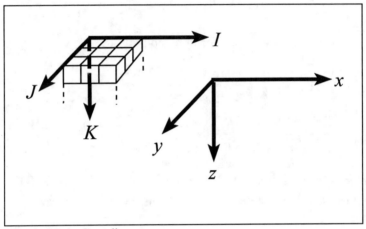

Figure 23-1. Coordinate system.

Chapter 24

Rock-Fluid Interaction Data

Some of the most critical data in terms of their effect on simulator performance are the relative permeability and capillary pressure curves. They model the interaction between reservoir rock and fluids. Unfortunately, relative permeability curves are often among the missing or poorer quality data. Relative permeability data are affected significantly by alterations in wettability conditions in the core. Ideally, the relative permeability data should be measured in the laboratory under the same conditions of wettability that exist in the reservoir. One method of approaching this ideal is to use preserved, "native state" core samples. These are cores that are drilled using crude oil or a special coring fluid designed to minimize wettability alterations. The cores are sealed at the well site to minimize exposure to oxygen or drying and then preserved until ready to undergo flow testing in the laboratory. However, this process is expensive and most relative permeability data are obtained on restored state cores in the laboratory.

24.1 Three-phase Relative Permeability

In principle, three-phase relative permeability should be used when oil, water, and gas are flowing simultaneously. As a practical matter, the difficulty of accurately measuring three-phase relative permeabilities often makes their use meaningless. It is often sufficient to work with the two-phase relative permeability curves only.

Despite their shortcomings, it may be of interest to perform a simulation using a set of three-phase relative permeability curves. For this case, BOAST4D contains an option for computing a three-phase oil relative permeability curve using water-oil and gas-oil relative permeability curves. As with most calculations of this type, we assume:

a.   The water relative permeability curve (k_{rw}) obtained for a water-oil system depends only on water saturation, and

b.   The gas relative permeability curve (k_{rg}) obtained for a gas-oil system depends only on gas saturation.

The validity of these assumptions depends on such factors as wettability and degree of consolidation. Given the above assumptions, k_{rw} and k_{rg} for water-oil and gas-oil systems, respectively, are also valid for a water-gas-oil system. The three-phase oil relative permeability k_{ro3} is calculated as

$$k_{ro3} = \frac{(k_{row} + k_{rw})(k_{rog} + k_{rg})}{k_{row}^*} - (k_{rw} + k_{rg}) \qquad (24.1)$$

where

k_{row} =   oil relative permeability for water-oil system,

k_{rog} =   oil relative permeability for gas-oil system,

k_{row}^* =   oil relative permeability for water-oil system evaluated at the oil saturation corresponding to irreducible water saturation.

Equation (24.1) is based on the work by Stone [1973], and it corresponds to Model II of Dietrich and Bondor [1976].

When the three-phase calculation is activated, the user must be sure the input water-oil and gas-oil relative permeability curves are realistic. For example, if we write irreducible water saturation as S_{wr}, the relative permeability constraint $k_{row}(1 - S_{wr}) = k_{rog}(S_o + S_w = 1.0)$ must be satisfied since $S_g = 0$ in both cases.

References

Dietrich, J.K. and P.L. Bondor (1976): "Three-Phase Oil Relative Permeability Models," SPE Paper 6044, *Proceedings of 51st Fall Technical Conference and Exhibition of Society of Petrolem Engineers and AIME,* New Orleans, LA, Oct. 3-6.

Stone, H.L. (Oct.-Dec. 1973); "Estimation of Three-Phase Relative Permeability and Residual Oil Data," *Journal of Canadian Petroleum Technology,* pp. 53ff.

Chapter 25

PVT (Pressure Dependent) Data

25.1 General Comments

Laboratory reservoir fluid analyses generally provide data from both a differential liberation experiment and a flash experiment approximating field separator conditions. The differential and flash liberation data can be significantly different for some oils. The actual behavior of the production process is some combination of the differential and flash processes. The assumption normally made in preparing PVT data for use in a black oil simulator is that the differential liberation data represent the process occurring in the reservoir and the flash data represent production to stock tank conditions. Thus, for use in the simulator, the differential liberation data should be corrected to flash values at field separation conditions. This procedure is described in the literature [Amyx, et al., 1960; Moses, 1986] and is summarized below.

Physical property data obtained from a testing laboratory for a black oil system will generally be a differential liberation study coupled with a separator study. Most reservoir simulators require that these data be converted to flash form so that the effects of the surface separation facility are included. Conversion of the data is restricted to oil formation volume factor and solution gas-oil ratio data. If the separator B_o and R_{so} are known, the conversion equations are:

$$B_o(p) = B_{od}(p) \; \frac{B_{ofbp}}{B_{odbp}}$$

and

$$R_{so}(p) = R_{sofbp} - \left(R_{sodbp} - R_{sod}(p)\right) \frac{B_{ofbp}}{B_{odbp}}$$

where subscripts are defined as:

d = differential liberation data

f = flash data

bp = bubble point

Reservoir fluid properties (PVT data) include fluid viscosities, densities, formation volume factors, gas solubilities, etc. These data are usually obtained by laboratory analyses applied to fluid samples taken from the reservoir. They are sketched in Figures 25-1 and 25-2. Often the PVT data are not known over as wide a range of pressures as would be desirable for a computer run. When this occurs, the fluid data base can be broadened by complementing the laboratory data with correlations and by extrapolating the laboratory-measured data.

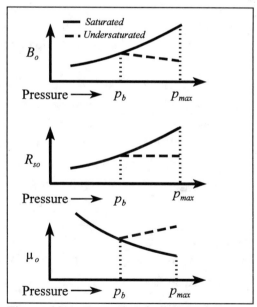

Figure 25-1. Example of oil PVT data.

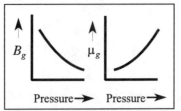

Figure 25-2. Example of gas PVT data.

25.2 Extrapolating Saturated Curves

Guidelines for extrapolating PVT data to pressures above the measured saturation pressure are presented below.

1. The B_g versus pressure curve is strongly non-linear and an extrapolation of this curve to small B_g values at high pressures can result in errors. For most natural gases, the relationship $1/B_g$ versus pressure will be very nearly linear, especially at moderate to high pressures. Plotting $1/B_g$ versus pressure and extrapolating to PMAX should provide more realistic values of B_g at higher pressures. Interpolating B_g using $1/B_g$ versus pressure substantially improves material balance.

2. Once the B_g versus P curve is fixed, R_{so} versus P and B_o versus P curves must be extrapolated so as to avoid a negative oil compressibility being calculated over any pressure increment. To ensure that negative oil compressibilities will not be calculated by the program, the following test should be used. For any pressure increment P_1 to P_2, where $P_2 > P_1$. the following relationship should hold:

$$0 \le -\left(B_{o2} - B_{o1}\right) + \frac{B_{g2}(R_{so2} - R_{so1})}{5.615} \qquad (25.1)$$

where the units of B_o, B_g, and R_{so} are RB/STB, RCF/SCF, and SCF/STB, respectively. Note that this test applies only to the saturated oil PVT data.

3.   The above concepts also apply to the water PVT data. However, for most simulations, it can be assumed that R_{sw} = 0.0, thus - $\Delta B_w /B_w \Delta P$ approximates water compressibility.

25.3 Gas PVT Correlation Option

Basic Gas Properties:

Following Govier [1975], real gas Z-factors are computed using the Dranchuk, et al. [1974] representation of the Standing-Katz Z-factor charts [1942]. This representation employs the Benedict-Webb-Rubin [1940] eight parameter equation of state to express the Z-factor as a function of pseudo-critical temperature T_r and pseudo-critical pressure P_r, thus

$$Z = Z(P_r, T_r). \tag{25.2}$$

Once Z is known, the gas formation volume factor is easily determined for a given temperature and pressure using the real gas law.

The isothermal gas compressibility c_g is obtained from Eq. (25.2) as

$$c_g = \frac{1}{P_c} \left[\frac{1}{P_r} - \frac{1}{Z} \left(\frac{\partial Z}{\partial P_r} \right)_{T_r} \right] \tag{25.3}$$

where P_c is the critical pressure (psia).

Real gas viscosities are computed using the method described in Govier [1975]. This method is a computerized version of the Carr, Kobayashi and Burrows [1954] hydrocarbon gas viscosity determination procedure.

Pseudo-Pressure Calculations:

Pseudo-pressures are defined by

$$\psi(P) = 2 \int_{P_o}^{P} \frac{P'}{\mu_g Z} \, dP' \tag{25.4}$$

where

$P' =$ dummy integration variable with pressure units (psia)

$P_o =$ reference pressure = 14.7 psia

$P =$ specified pressure (psia)

$\mu_g =$ gas viscosity (cp)

$Z =$ gas compressibility factor

The pseudo-pressure $\psi(P)$ is often written as $m(P)$. Since μ_g and Z depend on P', evaluation of Eq. (25.4) is accomplished by numerical integration using the trapezoidal rule and a user-specified pressure increment $\Delta P' \approx dP'$.

Gas Property Description:

Four different gas property descriptions may be specified. Their descriptions and control parameter (KODEA) values follow:

KODEA	GAS DESCRIPTION
1	Sweet gas: input 12 component mole fractions as 0. 0. 0. 1. 0. 0. 0. 0. 0. 0. 0. 0.
2	Sour gas: input 12 component mole fractions in the order y_1 y_2 y_3 y_4 0. 0. 0. 0. 0. 0. 0. 0. where y_1 = mole fraction of H_2S, y_2 = mole fraction of CO_2 y_3 = mole fraction of N_2, and $y_4 = 1 - (y_1 + y_2 + y_3)$.
3	Sweet or sour gas with the following 12 component mole fractions read in the following order: H_2S, CO_2, N_2, C_1, C_2, C_3, iC_4, nC_4, iC_5, nC_5, C_6, C_{7+}. The sum of the mole fractions should equal one.
4	Same as KODEA = 3 but also read critical pressure, critical temperature, and molecular weight of C_{7+}.

Correlation Range Limits:

The following range limits apply to correlations used in calculating gas Z-factors, compressibilities and viscosities:

$$1.05 < \frac{T}{T_c} < 3.0$$

$$0.01 < \frac{P}{P_c} < 15.0$$

$$0.55 < SPG < 1.5$$

$$40 < T < 400$$

where

T_c = pseudo-critical temperature (°R)
P_c = pseudo-critical pressure (psia)
T = temperature (°R)
P = pressure (psia)
SPG = gas specific gravity

No values of T, P, or SPG should be used that exceed the above correlation ranges. If the range limit is exceeded, a fatal error will occur.

References

Amyx, J.W., D.M. Bass, and R.L. Whiting (1960); **Petroleum Reservoir Engineering**, New York: McGraw-Hill.

Benedict, M., G.B. Webb, and L.C. Rubin (1940): *Journal of Chemical Physics,* Volume 8, 334.

Carr, N.L., R. Kobayashi, and D.B. Burrows (1954): "Viscosity of Hydrocarbon Gases Under Pressure," *Transactions of the AIME,* Volume 201, pp 264-272.

Dranchuk, P.M., R.A. Purvis, and D.B. Robinson (1974): "Computer Calculation of Natural Gas Compressibility Factors Using the Standing and Katz Correlations," *Institute of Petroleum Technology,* IP-74-008.

Govier, G.W., Editor (1978): **Theory and Practice of the Testing of Gas Wells**, Calgary: Energy Resources Conservation Board.

Moses, P.L. (July 1986): "Engineering Applications of Phase Behavior of Crude Oil and Condensate Systems," *Journal of Petroleum Technology,* pp. 715-723

Standing, M.B. and D.L. Katz (1942): "Density of Natural Gases," *Transactions of the AIME,* Volume 146, 140.

Chapter 26

Aquifer Models

A reservoir-aquifer system can be modeled using small grid blocks to define the reservoir and increasingly larger grid blocks to define the aquifer. This approach has the advantage of providing a numerically uniform analysis of the reservoir-aquifer system, but it has the disadvantage of requiring more computer storage and computing time because additional grid blocks are used to model the aquifer. A more time- and cost-effective means of representing an aquifer is to represent aquifer influx with an analytic model. Three models are available as options in BOAST4D.

26.1 Pot Aquifer

Aquifer influx is calculated assuming the aquifer is both small and bounded. The pot aquifer influx rate q_{wp} is dependent on the pressure change over a time step for a specified grid block:

$$q_{wp} = -\left[POT \; \frac{(P^n - P^{n+1})}{\Delta t} \right]; \; POT \geq 0 \qquad (26.1)$$

where P^n, Δt^n are grid block pressure and time step at the present time level n; P^{n+1}, Δt^{n+1} are grid block pressure and time step at the future time level $n + 1$; and POT is the pot aquifer coefficient. The minus sign preceding the bracketed term indicates water is entering the block when $P^n > P^{n+1}$.

26.2 Steady-state Aquifer

The steady-state aquifer model is based on Schilthuis's assumption that the water influx rate q_{wss} is proportional to the pressure difference between the aquifer and the hydrocarbon reservoir. It is further assumed that the aquifer is sufficiently large that it experiences no net pressure change throughout the producing life of the reservoir. With these assumptions, BOAST4D computes steady-state aquifer influx into a specified grid block as

$$q_{wss} = -\left[\text{SSAQ} \ (P^0 - P^{n+1})\right]; \ \text{SSAQ} \geq 0 \qquad (26.2)$$

where P^{n+1} is the grid block pressure at the future time level $n + 1$; P^0 is the initial grid block pressure; and SSAQ is the proportionality constant. The minus sign preceding the bracketed term indicates water is entering the block when $P^0 > P^{n+1}$.

26.3 Carter-Tracy Aquifer

The Carter-Tracy [1960] modification of the Hurst-van Everdingen [1949] unsteady-state aquifer influx calculation is available in BOAST4D. The Carter-Tracy aquifer influx rate q_{wct} for a specified grid block is

$$q_{wct} = -[A - B(P^{n+1} - P^n)] \qquad (26.3)$$

where P^n, P^{n+1} are grid block pressures at time levels n and $n + 1$, respectively. The coefficients A and B are given by

$$A = K_t \left[\frac{\beta(P^0 - P^n) - W_e^n P_{td}^{'n+1}}{\text{DENOM}}\right] \qquad (26.4)$$

with

$$B = K_t \frac{\beta}{\text{DENOM}} \qquad (26.5)$$

$$\text{DENOM} = P_{tD}^{n+1} - t_D^n P_{tD}^{'n+1} \qquad (26.6)$$

$$P_{tD}^{'n+1} = \left[\frac{dP_{tD}}{dt_D} \right]^{n+1} \tag{26.7}$$

$$K_t = 0.00633 \frac{k}{(\phi \mu c r_e^2)} = \text{AQPAR 1} \tag{26.8}$$

$$\beta = 2\pi \phi h c r_e^2 s = \text{AQPAR 2} \tag{26.9}$$

and

$$c = c_r + c_w \tag{26.10}$$

The quantities t_D and P_{tD} are dimensionless time and pressure, respectively, with

$$t_D = K_t t$$

and P_{tD} is the Carter-Tracy influence function for the constant terminal rate case. The functions P_{tD} and P'_{tD} are numerically represented by regression equations [Fanchi, 1985]. All remaining parameters are defined as follows:

c_r	=	rock compressibility (psi^{-1})
c_w	=	water compressibility (psi^{-1})
h	=	aquifer net thickness (ft)
k	=	aquifer permeability (md)
r_e	=	external aquifer radius (ft)
r_w	=	external reservoir radius (ft)
s	=	$\theta/360°$ where θ is the angle of aquifer/reservoir interface
W_e^n	=	cumulative water influx at time level n, SCF
μ	=	aquifer water viscosity (cp)
ϕ	=	aquifer porosity

References

Carter, R.D. and G.W. Tracy (1960) "An Improved Method for Calculating Water Influx," *Transactions of the AIME*, Vol. 219, pp. 415-417.

van Everdingen, A.F. and W. Hurst (1949): "The Application of the Laplace Transformation to Flow Problems in Reservoirs," *Transactions of the AIME*, Vol. 186, pp. 305-324.

Fanchi, J.R. (June 1985): "Analytical Representation of the van Everdingen-Hurst Aquifer Influence Functions for Reservoir Simulation," *Society of Petroleum Engineers Journal*, pp. 405-406.

Chapter 27

Initialization

It is important when making cross-section or 3D runs that the pressures in the model are correctly initialized. If not, phase potential differences due to gravity terms could cause fluid migration even though no wells are active. Consequently, a simple pressure initialization algorithm is used in BOAST4D. It is reviewed below along with an option to correct pressures to a user-specified datum and an option to initialize saturations using gravity segregation.

27.1 Pressure Initialization

Consider a grid block that may have a gas-oil contact and a water-oil contact as in Figure 27-1.

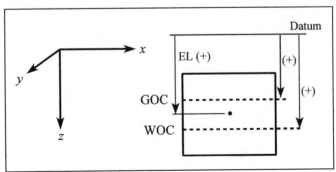

Figure 27-1. Depths for pressure initialization algorithm.

We assume the pressure in the grid block at model location (i, j, k) is dominated by the density of the phase at the block midpoint and that there are no transition zones between different phases initially. The pressure and depth at the gas-oil contact are PGOC and GOC, respectively. Similarly, for the water-oil contact we have PWOC and WOC.

The initial pressure assigned to the grid block in Figure 27-1 is determined by the depth of the node (midpoint) relative to the respective contact elevations.

Let us define the depth of the block midpoint from datum as EL_{ijk}. With this definition, the pressure in the block is given by the following algorithm:

a. If $EL_{ijk} <$ GOC then

$$\rho_g = \rho_{gsc}/B_g \text{ and } P_{ijk} = \text{PGOC} + \rho_g \, (EL_{ijk} - \text{GOC})/144$$

b. If $EL_{ijk} >$ WOC then

$$\rho_w = (\rho_{wsc} + R_{sw} \cdot \rho_{gsc})/B_w \text{ and}$$

$$P_{ijk} = \text{PWOC} + \rho_w \, (EL_{ijk} - \text{GOC})/144$$

c. If GOC $\leq EL_{ijk} \leq$ WOC then

$$\rho_o = (\rho_{osc} + R_{so} \cdot \rho_{gsc})/B_o \text{ and}$$

$$P_{ijk} = \text{PWOC} + \rho_o \, (EL_{ijk} - \text{GOC})/144$$

The above algorithm should be reasonable for systems with initial transition zones that are small relative to the total thickness of the formation.

27.2 Pressure Corrected to Datum

Pressure P(I, J, K) of grid block I, J, K with mid-point elevation EL(I, J, K) may be corrected to a datum depth PDATUM by specifying a pressure gradient GRAD. The pressure at datum is given by PDAT(I, J, K) = P(I, J, K) + (PDATUM - EL(I, J, K))*GRAD.

27.3 Gravity Segregated Saturation Initialization

A simple model of a gravity segregated saturation distribution is calculated when KSI = 1. For depths increasing downward, we calculate elevations and thicknesses using the geometry shown in Figure 27-1 as follows:

$$\text{Block BOT} \quad = \quad \text{EL} + 0.5 *\text{DZ}$$
$$\text{Block THICK} \quad = \quad \text{DZ}$$
$$\text{Block TOP} \quad = \quad \text{BOT - THICK}$$

Water zone thickness

$$\text{WTHICK} \quad = \quad \text{BOT - WOC}$$

Gas zone thickness

$$\text{GTHICK} \quad = \quad \text{GOC - TOP}$$

The user must specify the initial oil saturation (SOI) for an oil-water system and the initial gas saturation (SGI) for a water-gas system. Given the initial saturations SOI and SGI, the following algorithm is applied [Fanchi, 1986].

Case			
Case 1	GOC ___ TOP ___ BOT ___ WOC ___	$S_g = 0$ $S_o = SOI$ $S_w = 1 - SOI$	
Case 2	TOP ___ GOC ___ $\}f_g$ WOC ___ BOT ___ $\}f_w$	$f_g = \dfrac{\text{GTHICK}}{\text{THICK}}$ $f_w = \dfrac{\text{WTHICK}}{\text{THICK}}$ $S_g = f_g *SGI$ $S_o = (1 - f_g - f_w) *SOI$ $S_w = 1 - S_o - S_g$	If $S_o < S_{or}$, then $S_o = 0$ $S_g = \dfrac{f_g *SGI}{(f_g + f_w)}$ $S_w = 1 - S_g$
Case 3	TOP ___ GOC ___ BOT ___ WOC ___ $\Big\}f$	$f = 1 - \dfrac{\text{GTHICK}}{\text{THICK}}$ $S_o = 1 - SOI*f$ $S_g = (1 - f) *SGI$ $S_w = 1 - S_o - S_g$	If $S_o < S_{or}$, then $S_o = 0$ $S_w = 1 - SGI$ $S_g = SGI$

Case 4	GOC ___ TOP ___ $\left.\begin{array}{l}\\ \\ \\ WOC ___ \\ BOT ___ \end{array}\right\}f$	$f = 1 - \dfrac{WTHICK}{THICK}$ $S_g = 0$ $S_w = 1 - SOI*f$ $S_o = SOI*f$	If $S_o < S_{or}$, then $S_o = 0$ $S_w = 1$
Case 5	TOP ___ BOT ___ GOC ___ WOC ___	$S_o = 0$ $S_w = 1 - SGI$ $S_g = SGI$	
Case 6	GOC ___ WOC ___ TOP ___ BOT ___	$S_o = S_g = 0$ $S_w = 1$	

Water saturation is calculated as $S_w = 1 - S_o - S_g$ in all cases. Cases 2 through 4 require the user to enter residual oil saturation S_{or}.

References

Fanchi, J.R. (1986): "BOAST-DRC: Black Oil and Condensate Reservoir Simulation on an IBM-PC," Paper SPE 15297, *Proceedings from Symposium on Petroleum Industry Applications of Microcomputers of SPE,* Silver Creek, CO, June 18-20.

Chapter 28

Well Models

The well models contained in BOAST4D are described in this chapter. User specified parameters for controlling these well models are defined in Chapter 20.

28.1 Rate Constraint Representation

Case 1: Oil Production Rate Q_o Specified

In this representation, rates may be specified for injectors or producers. We assume the well may be completed in a total of K connections, and the production rates for each connection k for a specified oil rate are:

Oil

$$Q_{ok} = Q_o \frac{\left[(PID) \dfrac{\lambda_o}{B_o} \right]_k}{\displaystyle\sum_{k=1}^{K} \left[(PID) \dfrac{\lambda_o}{B_o} \right]_k} \qquad (28.1)$$

Water

$$Q_{wk} = Q_{ok} \left(\frac{\lambda_w / B_w}{\lambda_o / B_o} \right)_k \qquad (28.2)$$

Gas

$$Q_{gk} = \left(\frac{\lambda_g/B_g}{\lambda_o/B_o} \right)_k Q_{ok} + (R_{so})_k Q_{ok} + (R_{sw})_k Q_{wk} \tag{28.3}$$

where λ_ℓ is the fluid mobility of phase ℓ and PID is the well productivity index. For a more detailed discussion of PID, see Chapter 29. Notice that a PID may be specified for each connection. This capability lets the BOAST4D user take into account permeability contrast.

Case 2: Water Production Rate Q_w Specified

Assuming the well may be completed in K connections, the production rates of connection k for a specified water rate are:

Water

$$Q_{wk} = Q_w \frac{\left[(PID)\lambda_w/B_w \right]_k}{\displaystyle\sum_{k=1}^{K} \left[(PID)\lambda_w/B_w \right]_k} \tag{28.4}$$

Oil

$$Q_{ok} = Q_{wk} \left(\frac{\lambda_o/B_o}{\lambda_w/B_w} \right)_k \tag{28.5}$$

Gas

$$Q_{gk} = \left(\frac{\lambda_g/B_g}{\lambda_w/B_w} \right)_k Q_{wk} + (R_{sw})_k Q_{wk} + (R_{so})_k Q_{ok} \tag{28.6}$$

Case 3: Gas Production Rate Q_g Specified

Assuming the well may be completed in K connections, the production rates of connection k for a specified gas rate are:

Gas

$$Q_{gk} = Q_g \frac{\left[(PID)\lambda_g/B_g \right]_k}{\displaystyle\sum_{k=1}^{K} \left[(PID)\lambda_g/B_g \right]_k} \tag{28.7}$$

Oil

$$Q_{ok} = Q_{gk} \left(\frac{\lambda_o/B_o}{\lambda_g/B_g} \right)_k \tag{28.8}$$

Water

$$Q_{wk} = Q_{gk} \left(\frac{\lambda_w/B_w}{\lambda_g/B_g} \right)_k \tag{28.9}$$

Solution gas in both oil and water is neglected when a gas production rate is specified. This is a reasonable assumption for wells producing primarily free gas.

Case 4: Total Production Rate Specified

When the total reservoir voidage rate Q_T is specified, we first compute the phase mobility ratio for all connections:

Oil Mobility Ratio

$$\alpha_{oT} = \sum_{k=1}^{K} \left(\frac{\lambda_o}{\lambda_o + \lambda_w + \lambda_g} \right)_k \tag{28.10}$$

Water Mobility Ratio

$$\alpha_{wT} = \sum_{k=1}^{K} \left(\frac{\lambda_w}{\lambda_o + \lambda_w + \lambda_g} \right)_k \tag{28.11}$$

Gas Mobility Ratio

$$\alpha_{gT} = \sum_{k=1}^{K} \left(\frac{\lambda_g}{\lambda_o + \lambda_w + \lambda_g} \right)_k \tag{28.12}$$

We now compute the total oil rate

$$Q_o = \left(\frac{\alpha_{oT}}{\alpha_{oT} + \alpha_{wT} + \alpha_{gT}} \right) \frac{Q_T}{\overline{B}_o} \tag{28.13}$$

where

$$\overline{B}_o = \frac{1}{K} \sum_{k=1}^{K} (B_o)_k \tag{28.14}$$

is the average oil formation volume factor for all connections in which the well is completed. Given Eq. (28.13), we simply proceed as in Eqs. (28.1) through (28.3) above.

Case 5: Injection Rate Specified

If the well is a water or gas injector, the user must specify the total water or gas injection rates Q_w or Q_g, respectively, and a well injectivity index (WI) for each connection. The injection rate for each connection is then allocated as follows:

Water Injection Rate

$$Q_{wk} = Q_w \frac{\left[\mathrm{WI}(\lambda_o + \lambda_w + \lambda_g) \right]_k}{\sum\limits_{k=1}^{K} \left[\mathrm{WI}(\lambda_o + \lambda_w + \lambda_g) \right]_k} \qquad (28.15)$$

Gas Injection Rate

$$Q_{gk} = Q_g \frac{\left[\mathrm{WI}(\lambda_o + \lambda_w + \lambda_g) \right]_k}{\sum\limits_{k=1}^{K} \left[\mathrm{WI}(\lambda_o + \lambda_w + \lambda_g) \right]_k} \qquad (28.16)$$

It is important to note that allocation of injection fluids is based on total mobilities, and not just injected fluid mobility. This is necessary for the following reason: If an injector is placed in a block where the relative permeability to the injection fluid is zero, then the simulator using injection fluid mobility only would prohibit fluid injection even though a real well would allow fluid injection. A common example would be water injection into a block containing oil and irreducible water. To avoid the unrealistic result of no fluid injection, we assume the total mobility of the block should be used. For most cases, the error of this method will only persist for a few time steps because, in time, the mobile fluid saturation in the block will be dominated by the injected fluid.

28.2 Explicit Pressure Constraint Representation

Case 1: Oil and/or Water Production Wells

We assume that flowing bottomhole pressures (PWF) and well PIDs are specified for a pressure constrained well. The oil and water rates in STB/D for connection k are given by

$$Q_{ok} = \left[PID \frac{\lambda_o}{B_o} \right]_k^n (P^n - PWF)_k \qquad (28.17)$$

and

$$Q_{wk} = \left[PID \frac{\lambda_w}{B_w} \right]_k^n (P^n - PWF)_k \qquad (28.18)$$

where the explicit pressure P^n is used. If $P^n < PWF$, the well is shut in. When $P^n > PWF$, Q_{ok} and Q_{wk} are calculated and then substituted into Eq. (28.3) to find Q_{gk}.

Case 2: Gas Production Well

The laminar-inertial-turbulent (LIT) method may be used to represent a gas production well. The LIT method entails fitting gas well test data to the equation

$$\Delta \psi = aQ_g + bQ_g^2 = \psi_R - \psi_{wf} \qquad (28.19)$$

where

ψ_R = pseudo-pressure corresponding to shut-in pressure P_R (psia²/cp)

ψ_{wf} = pseudo-pressure corresponding to a specified well flowing pressure P_{wf} (psia²/cp)

aQ_g = laminar flow

bQ_g^2 = inertial and turbulent flow

BOAST4D employs user specified values of a, b, P_{wf}, and a table of pseudo-pressure versus pressure values to compute the total gas well production rate as

$$Q_g = \frac{-a + \sqrt{a^2 + 4b\Delta\psi}}{2b} \tag{28.20}$$

where ψ_R is the pseudo-pressure corresponding to the nodal pressure P^n. Rates for each phase in connection k are computed by mobility allocation as shown in Eqs. (28.7) through (28.9).

Case 3: Injection Wells

The injection rate for a water or gas injection well is computed from

$$Q_{pk} = \left[PID\left(\frac{\lambda_o + \lambda_w + \lambda_g}{B_p} \right) \right]^n_k (P^n - PWF)_k \tag{28.21}$$

where the subscript p denotes water or gas, and PID = WI. Fluid injection occurs when $P^n <$ PWF. If $P^n >$ PWF, the injection well is shut in. Also note that total mobility is used for the injection well rate calculation. The reason for this was discussed in the first section of this chapter.

28.3 GOR/WOR Constraints

Maximum gas-oil and water-oil ratios (GORMAX, WORMAX respectively) are input by the user and apply to every oil production well. GOR for a well is defined as total gas production divided by total oil production for all active well completion intervals. If GOR for the well exceeds GORMAX, then the completion interval (connection) with the highest GOR will be shut in. If more than one connection has the same maximum GOR, the shallowest connection will be shut in first. The procedure is repeated until GOR is less than GORMAX or until the well is shut in.

The ratio WOR is defined as total water production divided by total oil production for all active well completion intervals. If WOR for the well exceeds WORMAX, then the completion interval (connection) with the highest WOR will be shut in. If more than one connection has the same maximum WOR, the deepest connection will be shut in first. The procedure is repeated until WOR is less than WORMAX or until the well is shut in.

28.4 Fluid Withdrawal Constraints

Fluid withdrawal from explicit pressure controlled production wells can be constrained as follows:

a.   A minimum oil production rate can be specified;

b.   A maximum oil production rate can be specified; and

c.   A maximum liquid (water plus oil) withdrawal rate can be specified.

A positive value of QO for a pressure controlled production well is used as the minimum allowed oil production rate. If the calculated oil production rate drops below the minimum allowed value, the well is shut in.

A positive value of QW for a pressure controlled production well is used as the maximum allowed oil production rate. If the calculated oil production rate exceeds the maximum allowed value, calculated production will be reduced to the allowed value. Production from each connection is proportionally reduced by the ratio of allowed to calculated oil production rates.

A positive value of QT for a pressure controlled production well is used as the maximum allowed liquid withdrawal rate. If the sum of oil and water production exceeds the maximum allowed value, calculated production is reduced to the allowed value. The reduction is made by multiplying production from each connection by the ratio of allowed to calculated liquid withdrawal rates. **IMPORTANT:** When used to control total liquid withdrawal, the units of QT are STB/Day.

28.5 Fluid Injection Constraints

Fluid injection into explicit pressure controlled injection wells can be constrained as follows:

a.   A maximum water injection rate can be specified; and

b.   A maximum gas injection rate can be specified.

A negative value of QW for a pressure controlled water injection well is used as the maximum allowed water injection rate. If the calculated water injection rate exceeds the allowed value, calculated water injection will be reduced to the allowed value. Water injection into each connection is proportionally reduced by the ratio of allowed to calculated water injection rates.

A negative value of QG for a pressure controlled gas injection well is used as the maximum allowed gas injection rate in direct analogy to the water injection rate constraint described previously.

Chapter 29

Estimating Well Flow Index (PID)

29.1 Productivity Index

Productivity index (PI) is defined as the ratio of rate Q to pressure drop ΔP, or PI $= Q/\Delta P$, where $\Delta P = P_e - P_w$, P_e = average reservoir pressure, and P_w = wellbore bottomhole pressure BHP. From Darcy's Law for radial oil flow we can write PI as

$$ \text{PI} = \frac{Q_o}{\Delta P} = \frac{0.00708\,K_e\,h_{net}}{\mu_o B_o\left[\ell n\left(r_e/r_w\right) + S\right]} \tag{29.1}$$

The meaning and units of all terms are given as follows:

μ_o	=	oil viscosity (cp)
B_o	=	oil FVF (RB/STB)
r_e	=	drainage radius (ft)
r_w	=	wellbore radius (ft)
S	=	skin
K_e	=	effective permeability (md) $= k_{ro}\,K_{abs}$
k_{ro}	=	relative permeability to oil
K_{abs}	=	absolute permeability (md)
h_{net}	=	net thickness (ft)
Q_o	=	oil rate (STB/D)

Some of the terms in Eq. (29.1) depend on time-varying pressure and saturation, while other factors change relatively slowly or are constant with respect to time. We separate these terms to obtain

$$PI = \frac{k_{ro}}{\mu_o B_o} \, PID$$

where the quasi-stationary factors are collected in the PID term, that is,

$$PID = \frac{0.00708 K_{abs} h_{net}}{\ell n(r_e/r_w) + S}$$

The BOAST4D user is expected to provide a PID for each well connection.

29.2 Vertical Wells

A value of the connection flow index PID for a vertical well can be estimated from a formula derived by Peaceman [1978]:

$$PID_k = \left[\frac{0.00708 Kh}{ln\left(\dfrac{r_e}{r_w}\right) + S} \right]_k \tag{29.2}$$

where

$$r_e \approx r_o = 0.14(\Delta x^2 + \Delta y^2)^{\frac{1}{2}}$$

for an isotropic system. With respect to permeability, an isotropic system is a system in which x direction and y direction permeabilities are equal, ($K_x = K_y$). For a square well block in an isotropic system, $\Delta x = \Delta y$ and $r_o \approx 0.2 \, \Delta x$. The subscript k in Eq. (29.2) denotes the kth connection. For a well in a rectangular grid block and an anisotropic system (that is, $K_x \neq K_y$), well PID is estimated using an effective permeability

$$K = \sqrt{K_x K_y}$$

and an equivalent well block radius

$$r_e \approx r_o = 0.28 \frac{[(K_y/K_x)^{\frac{1}{2}} \Delta x^2 + (K_x/K_y)^{\frac{1}{2}} \Delta y^2]^{\frac{1}{2}}}{(K_y/K_x)^{\frac{1}{4}} + (K_x/K_y)^{\frac{1}{4}}}$$

The remaining parameters are defined as:

K = horizontal permeability of connection k (md)

h = thickness of connection k (ft)

r_w = wellbore radius (ft)

S = dimensionless skin factor

In principle, the well flow index can be related to measured values. In practice, however, the terms r_e, S, and $k_{ro}/\mu_o B_o$ are seldom well known, especially for a multiphase flowing well. As a matter of expediency, therefore, Eq. (29.2) is often used to compute an initial estimate of PID. This value can then be improved by adjusting it until the simulator computed well rates match the initial observed well rates.

29.3 Horizontal Wells

There are many ways to estimate connection flow index PID for a horizontal well [Joshi, 1991]. A PID value can be estimated for horizontal wells in a manner similar to that for vertical wells by using the Joshi formula

$$PID_k = \left[\frac{0.00708 Kh}{\ln\left(\dfrac{a + \sqrt{a^2 - \left(\dfrac{L}{2}\right)^2}}{L/2}\right) + \dfrac{h}{L}\ln\left(\dfrac{h}{2r_w}\right) + S} \right]_k \qquad (29.3)$$

where

$$a = \frac{L}{2}\left[0.5 + \sqrt{0.25 + \left(\frac{2r_{eh}}{L}\right)^4}\right]^{\frac{1}{2}}$$

The subscript k in Eq. (29.3) denotes the kth connection. The remaining parameters are defined as:

K = horizontal permeability of connection k (md)

h = thickness of connection k (ft)

L = horizontal well length (ft)

r_w = wellbore radius (ft)

r_{eh} = drainage radius of horizontal well (ft)

S = dimensionless skin factor

References

Joshi, S.D. (1991): **Horizontal Well Technology**, Tulsa, OK: PennWell Publishing Company.

Peaceman, D.W. (June 1978): "Interpretation of Well-Block Pressures in Numerical Reservoir Simulation," *Society of Petroleum Engineering Journal*, pp. 183-194. See also Peaceman, D.W. (June 1983): "Interpretation of Well-Block Pressures in Numerical Reservoir Simulation With Nonsquare Grid Blocks and Anisotropic Permeability," *Society of Petroleum Engineering Journal*, pp. 531-543.

Chapter 30

Reservoir Geophysics

Monitoring changes in the seismic characteristics of a reservoir as the reservoir is produced is the basis of 4D seismic monitoring [Anderson, et al., 1995; He, et al., 1996]. This is done in BOAST4D by calculating seismic attributes as a function of time. The seismic attributes are defined below.

30.1 Compressional and Shear Velocities

Seismic compressional velocity and shear velocity are calculated from the expressions [Schön, 1996; McQuillin, et al., 1984]:

$$V_P = \sqrt{\frac{K^* + \dfrac{4\mu^*}{3}}{\rho_B}}$$

and

$$V_S = \sqrt{\frac{\mu^*}{\rho_B}}$$

where

V_P = compressional velocity
V_S = shear velocity
K^* = effective bulk modulus
μ^* = effective shear modulus
ρ_B = effective bulk density = $(1-\phi)\rho_{ma} + \phi\rho_f$

ρ_{ma} = density of grains (solid matrix material)

ρ_f = fluid density = $\rho_o S_o + \rho_w S_w + \rho_g S_g$

ϕ = porosity

Gassman [1951] derived an expression for K^* from the theory of elasticity of porous media [Schön, 1996; McQuillin, et al., 1984]:

$$K^* = K_M + \frac{\left[1 - \dfrac{K_M}{K_G}\right]^2}{\dfrac{\phi}{K_F} + \dfrac{1-\phi}{K_G} - \dfrac{K_M}{K_G^2}}$$

where

K_M = bulk modulus of empty reservoir, that is, dry rock or porous matrix material

K_G = bulk modulus of grains (solid matrix material)

K_F = bulk modulus of fluid = $1/c_f$

c_f = fluid compressibility = $c_o S_o + c_w S_w + c_g S_g$

The BOAST4D user must enter data that cannot be calculated from traditional black oil simulator input data. In particular, the user must enter K_M, K_G, μ^*, and ρ_{ma}. The references give values that may be used if the data are not available from well logs such as shear wave logging tools, or laboratory measurements of parameters such as the dry frame Poisson's ratio.

30.2 Acoustic Impedance and Reflection Coefficients

Acoustic impedance Z is defined as

$$Z = \rho_B V_P$$

The reflection coefficient RC at the interface between two layers with acoustic impedances Z_1 and Z_2 is given by

$$RC = \frac{Z_2 - Z_1}{Z_2 + Z_1}$$

The transmission coefficient TC is 1 - RC.

References

Anderson, R.N. (1995): "Method Described for Using 4D Seismic to Track Reservoir Fluid Movement," *Oil & Gas Journal*, pp. 70-74, April 3.

Gassman, F. (1951): "Elastic Waves Through a Packing of Spheres," *Geophysics*, Volume 16, 673-685.

He, W., R.N. Anderson, L. Xu, A. Boulanger, B. Meadow, and R. Neal (1996): "4D Seismic Monitoring Grows as Production Tool," *Oil & Gas Journal*, pp. 41-46, May 20.

McQuillin, R., M. Bacon, and W. Barclay (1984): **An Introduction to Seismic Interpretation**, Houston: Gulf Publishing.

Schön, J.H. (1996): **Physical Properties of Rocks: Fundamentals and Principles of Petrophysics**, Volume 18, New York: Elsevier.

Chapter 31

Material Balance

Material balance is one measure of the numerical stability and accuracy of a simulator. The BOAST4D material balance calculation at time t is given by

$$\text{Material Balance} = \frac{\text{FIP}}{\text{OFIP} - \text{Prod} + \text{Inj}}$$

where

    FIP   = fluid in place at time t
    OFIP = original fluid in place
    Prod = cumulative fluid produced at time t
    Inj   = cumulative fluid injected at time t

Based on this definition, material balance should equal one in an idealized calculation. Actual simulator material balance may not equal one.

Material balance error reported by BOAST4D is calculated using the formula

$$\% \text{ Error} = \left\{ \frac{\text{FIP}}{\text{OFIP} - \text{Prod} + \text{Inj}} - 1 \right\} \times 100\%$$

Material balance can be a sensitive indicator of error. Material balance error is greatest in BOAST4D when a grid block undergoes a phase transsition, for example, when a grid block passes from single phase oil to two-phase oil and gas during a time step.

Material balance errors can be corrected by adding or subtracting enough fluid to reestablish an exact material balance [Nolen and Berry, 1973; Spillette, et al., 1986]. This material balance correction technique is equivalent to adding a source/sink term to the mass conservation equations for every grid block. These terms are not included in the BOAST4D formulation. The exercises in Part I show that the uncorrected formulation can be used with good accuracy in many practical situations.

References

Nolen, J.S. and D.W. Berry (1973): "Test of the Stability and Time Step Sensitivity of Semi-Implicit Reservoir Simulation Techniques," **Numerical Simulation**, SPE Reprint Series #11, Richardson, TX: Society of Petroleum Engineers.

Spillette, A.G., J.G. Hillestad, and H.L. Stone (1986): "A High-Stability Sequential Solution Approach to Reservoir Simulation," **Numerical Simulation II**, SPE Reprint Series #20, Richardson, TX: Society of Petroleum Engineers.

Chapter 32

Derivation of the Flow Equations

Many derivations of the oil, water, and gas fluid flow equations exist in the literature [for example, see Crichlow, 1977; Peaceman, 1977]. Consequently, only a brief discussion will be presented here. It closely follows the presentation originally published in Fanchi, et al. [1982].

32.1 Conservation of Mass

We begin by considering the flow of fluid into and out of a single reservoir block (Figure 32-1). Assume fluid flows into the block at x (J_x) and out of the block at $x + \Delta x$ ($J_{x+\Delta x}$). J denotes the fluid flux and is defined as the rate of flow of mass per unit cross-sectional area normal to the direction of flow, which is the x direction in the present case. By conservation of mass, we have the equality:

mass entering the block - mass leaving the block

= accumulation of mass in the block.

If the block has length Δx, width Δy, and depth Δz, then we can write the mass entering the block in a time interval Δt as

$$[(J_x)_x \Delta y \Delta z + (J_y)_y \Delta x \Delta z + (J_z)_z \Delta x \Delta y] \Delta t = \text{Mass in} \qquad (32.1)$$

where we have generalized to allow flux in the y and z directions as well. The notation $(J_x)_x$ denotes the x direction flux at location x, with analogous meanings for the remaining terms.

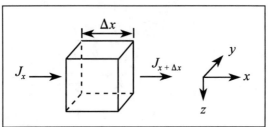

Figure 32-1. Reservoir block: the coordinate convention follows Sawyer and Mercer [1978].

Corresponding to mass entering is a term for mass exiting which has the form

$$[(J_x)_{x+\Delta x}\Delta y\Delta z + (J_y)_{y+\Delta y}\Delta x\Delta z + (J_z)_{z+\Delta z}\Delta x\Delta y]\Delta t$$

$$+ q\Delta x\Delta y\Delta z\Delta t = \text{Mass out}$$

(32.2)

We have added a source/sink term q which represents mass flow into (source) or out of (sink) a well. A producer is represented by $q > 0$, and an injector by $q < 0$.

Accumulation of mass in the block is the change in concentration of phase p (C_p) in the block over the time interval Δt. If the concentration C_p is defined as the total mass of phase p (oil, water, or gas) in the entire reservoir block divided by the block volume, then the accumulation term becomes

$$[(C_p)_{t+\Delta t} - (C_p)_t]\Delta x\Delta y\Delta z = \text{Mass accumulation}$$

(32.3)

Using Eqs. (32.1) through (32.2) in the mass conservation equality

Mass in - Mass out = Mass accumulation

gives

$$[(J_x)_x\Delta y\Delta z + (J_y)_y\Delta x\Delta z + (J_z)_z\Delta x\Delta y]\Delta t$$

$$- [(J_x)_{x+\Delta x}\Delta y\Delta z + (J_y)_{y+\Delta y}\Delta x\Delta z + (J_z)_{z+\Delta z}\Delta x\Delta y]\Delta t$$

$$- q\Delta x\Delta y\Delta z\Delta t = [(C_p)_{t+\Delta t} - (C_p)_t]\Delta x\Delta y\Delta z$$

(32.4)

Dividing Eq. (32.4) by $\Delta x\Delta y\Delta z\Delta t$ and rearranging gives

$$-\frac{(J_x)_{x+\Delta x} - (J_x)_x}{\Delta x} - \frac{(J_y)_{y+\Delta y} - (J_y)_y}{\Delta y}$$

$$-\frac{(J_z)_{z+\Delta z} - (J_z)_z}{\Delta z} - q = \frac{(C_p)_{t+\Delta t} - (C_p)_t}{\Delta t} \tag{32.5}$$

In the limit as Δx, Δy, Δz, and Δt go to zero, Eq. (32.5) becomes the continuity equation

$$-\frac{\partial J_x}{\partial x} - \frac{\partial J_y}{\partial y} - \frac{\partial J_z}{\partial z} - q = \frac{\partial C_p}{\partial t} \tag{32.6}$$

The oil, water, and gas phases each satisfy a mass conservation equation having the form of Eq. (32.6).

32.2 Flow Equations for Three-Phase Flow

The flow equations for an oil, water, and gas system are determined by specifying the fluxes and concentrations of the conservation equations for each of the three phases. A flux in a given direction can be written as the density of the fluid times its velocity in the given direction. Letting the subscripts o, w, and g denote oil, water, and gas, respectively, the fluxes become:

$$(\vec{J})_o = \frac{\rho_{osc}}{B_o}\vec{v}_o \tag{32.7}$$

$$(\vec{J})_w = \frac{\rho_{wsc}}{B_w}\vec{v}_w \tag{32.8}$$

$$(\vec{J})_g = \frac{\rho_{gsc}}{B_g}\vec{v}_g + \frac{R_{so}\rho_{gsc}}{B_o}\vec{v}_o + \frac{R_{sw}\rho_{gsc}}{B_w}\vec{v}_w \tag{32.9}$$

where R_{so} and R_{sw} are gas solubilities in SCF/STB, B_o, B_w, and B_g are formation volume factors in units of reservoir volume/standard volume, the subscript *sc* denotes standard conditions (usually 60°F and 14.7 psia in oilfield units), and ρ denotes fluid densities. The velocities $\vec{v}$ are assumed to be Darcy velocities and their x components are

$$v_{xo} = -K_x \lambda_o \frac{\partial}{\partial x} \left[P_o - \frac{\rho_o g z}{144 g_c} \right] \tag{32.10}$$

$$v_{xw} = -K_w \lambda_w \frac{\partial}{\partial x} \left[P_w - \frac{\rho_w g z}{144 g_c} \right] \tag{32.11}$$

$$v_{xg} = -K_x \lambda_g \frac{\partial}{\partial x} \left[P_g - \frac{\rho_g g z}{144 g_c} \right] \tag{32.12}$$

where g is the acceleration of gravity in ft/sec^2, and g_c is 32.174 ft/sec^2 (BOAST4D assumes $g = g_c$). These equations should be valid for describing fluid flow in porous media even if g and g_c change, such as on the Moon, Mars, or the space shuttle. Similar expressions can be written for the y and z components.

The phase mobility λ_p is defined as the ratio of the relative permeability to flow of the phase divided by its viscosity, thus

$$\lambda_p = k_{rp}/\mu_p \tag{32.13}$$

The phase densities are related to formation volume factors and gas solubilities by

$$\rho_o = \frac{1}{B_o} [\rho_{osc} + R_{so}\rho_{gsc}], \tag{32.14}$$

$$\rho_w = \frac{1}{B_w} [\rho_{wsc} + R_{sw}\rho_{gsc}], \tag{32.15}$$

$$\rho_g = \frac{\rho_{gsc}}{B_g}. \tag{32.16}$$

Besides fluxes, we also need concentrations. These are given by

$$C_o = \phi \rho_{osc} S_o / B_o, \tag{32.17}$$

$$C_w = \phi \rho_{wsc} S_w / B_w,$$ (32.18)

$$C_g = \phi \rho_{gsc} \left[\frac{S_g}{B_g} + R_{so} \frac{S_o}{B_o} + R_{sw} \frac{S_w}{B_w} \right]$$ (32.19)

where ϕ is the porosity and S_p is the saturation of phase p. The saturations satisfy the constraint

$$S_o + S_w + S_g = 1$$ (32.20)

Combining Eqs. (32.6), (32.7) through (32.9), and (32.17) through (32.19) gives a mass conservation equation for each phase:

Oil

$$-\left[\frac{\partial}{\partial x} \left(\frac{\rho_{osc}}{B_o} v_{xo} \right) + \frac{\partial}{\partial y} \left(\frac{\rho_{osc}}{B_o} v_{yo} \right) + \frac{\partial}{\partial z} \left(\frac{\rho_{osc}}{B_o} v_{zo} \right) \right]$$

$$- q_o = \frac{\partial}{\partial t} \left(\phi \rho_{osc} \frac{S_o}{B_o} \right)$$ (32.21)

Water

$$-\left[\frac{\partial}{\partial x} \left(\frac{\rho_{wsc}}{B_w} v_{xw} \right) + \frac{\partial}{\partial y} \left(\frac{\rho_{wsc}}{B_w} v_{yw} \right) + \frac{\partial}{\partial z} \left(\frac{\rho_{wsc}}{B_w} v_{zw} \right) \right]$$

$$- q_w = \frac{\partial}{\partial t} \left(\phi \rho_{wsc} \frac{S_w}{B_o} \right)$$ (32.22)

Gas

$$
-\frac{\partial}{\partial x}\left(\frac{\rho_{gsc}}{B_g}v_{xg} + \frac{R_{so}\rho_{gsc}}{B_o}v_{xo} + \frac{R_{sw}\rho_{gsc}}{B_w}v_{xw}\right)
$$

$$
-\frac{\partial}{\partial y}\left(\frac{\rho_{gsc}}{B_g}v_{yg} + \frac{R_{so}\rho_{gsc}}{B_o}v_{yo} + \frac{R_{sw}\rho_{gsc}}{B_w}v_{yw}\right)
$$

$$
-\frac{\partial}{\partial z}\left(\frac{\rho_{gsc}}{B_g}v_{zg} + \frac{R_{so}\rho_{gsc}}{B_o}v_{zo} + \frac{R_{sw}\rho_{gsc}}{B_w}v_{zw}\right) - q_g \tag{32.23}
$$

$$
= \frac{\partial}{\partial t}\left[\phi\rho_{gsc}\left(\frac{S_g}{B_g} + \frac{R_{so}S_o}{B_o} + \frac{R_{sw}S_w}{B_w}\right)\right]
$$

The densities at standard conditions are constants and can be divided out of the above equations. This reduces the equations to the following form:

Oil

$$
-\left[\frac{\partial}{\partial x}\left(\frac{v_{xo}}{B_o}\right) + \frac{\partial}{\partial y}\left(\frac{v_{yo}}{B_o}\right) + \frac{\partial}{\partial z}\left(\frac{v_{zo}}{B_o}\right)\right]
$$

$$
-\frac{q_o}{\rho_{osc}} = \frac{\partial}{\partial t}\left(\phi\frac{S_o}{B_o}\right) \tag{32.24}
$$

Water

$$
-\left[\frac{\partial}{\partial x}\left(\frac{v_{xw}}{B_w}\right) + \frac{\partial}{\partial y}\left(\frac{v_{yw}}{B_w}\right) + \frac{\partial}{\partial z}\left(\frac{v_{zw}}{B_w}\right)\right]
$$

$$
-\frac{q_w}{\rho_{wsc}} = \frac{\partial}{\partial t}\left(\phi\frac{S_w}{B_w}\right) \tag{32.25}
$$

Gas

$$
-\frac{\partial}{\partial x}\left(\frac{v_{xg}}{B_g} + \frac{R_{so}}{B_o}v_{xo} + \frac{R_{sw}}{B_w}v_{xw}\right)
$$

$$
-\frac{\partial}{\partial y}\left(\frac{v_{yg}}{B_g} + \frac{R_{so}}{B_o}v_{yo} + \frac{R_{sw}}{B_w}v_{yw}\right)
$$

$$
-\frac{\partial}{\partial z}\left(\frac{v_{zg}}{B_g} + \frac{R_{so}}{B_o}v_{zo} + \frac{R_{sw}}{B_w}v_{zw}\right)
$$

$$
-\frac{q_g}{\rho_{gsc}} = \frac{\partial}{\partial t}\left[\phi\left(\frac{S_g}{B_g} + R_{so}\frac{S_o}{B_o} + R_{sw}\frac{S_w}{B_w}\right)\right]
$$

(32.26)

Equations (32.10) through (32.16), (32.20), and (32.24) through (32.26) are the basic fluid flow equations which are numerically solved in a black oil simulator.

References

Crichlow, H.B. (1977): **Modern Reservoir Engineering – A Simulation Approach**, Englewood Cliffs, NJ: Prentice Hall.

Fanchi, J.R., K.J. Harpole, and S.W. Bujnowski (1982): "BOAST: A Three- Dimensional, Three-Phase Black Oil Applied Simulation Tool", 2 Volumes, U.S. Department of Energy, Bartlesville Energy Technology Center, OK.

Peaceman, D.W. (1977): **Fundamentals of Numerical Reservoir Simulation**, New York: Elsevier.

Sawyer, W.K. and J.C. Mercer (August 1978): "Applied Simulation Techniques for Energy Recovery," Department of Energy Report METC/RI-78/9, Morgantown, WV.

Chapter 33

The IMPES Formulation

The following section from Fanchi, et al. [1982] shows how these equations are recast in a form that is suitable for solution by a numerical technique. The numerical technique is based on the formulation originally presented by Sawyer and Mercer [1978].

33.1 Recasting the Flow Equations

A glance at Eqs. (32.24) through (32.26) illustrates the computational complexity of the basic three-dimensional, three-phase black oil simulator equations. Equivalent but much simpler appearing forms for the equations are

$$- \nabla \cdot \frac{\vec{v}_o}{B_o} - \frac{q_o}{\rho_{osc}} = \frac{\partial}{\partial t} \left(\phi \frac{S_o}{B_o} \right), \tag{33.1}$$

$$- \nabla \cdot \frac{\vec{v}_w}{B_w} - \frac{q_w}{\rho_{wsc}} = \frac{\partial}{\partial t} \left(\phi \frac{S_w}{B_w} \right), \tag{33.2}$$

and

$$- \nabla \cdot \left(\frac{\vec{v}_g}{B_g} + \frac{R_{so}}{B_o} \vec{v}_o + \frac{R_{sw}}{B_w} \vec{v}_w \right) - \frac{q_g}{\rho_{gsc}}$$

$$= \frac{\partial}{\partial t} \left[\phi \left(\frac{S_g}{B_g} + R_{so} \frac{S_o}{B_o} + R_{sw} \frac{S_w}{B_w} \right) \right] \tag{33.3}$$

where the symbol $\nabla \cdot \vec{v}$ is shorthand for

$$\nabla \cdot \vec{v} = \frac{\partial}{\partial x} v_x + \frac{\partial}{\partial y} v_y + \frac{\partial}{\partial z} v_z. \qquad (33.4)$$

For a review of vector analysis, see a reference such as Fanchi [1997].

The form of the Darcy velocities (Eqs. (32.10) through (32.12)) may be simplified by defining the potential Φ_p of phase p as

$$\Phi_p = P_p - \frac{\rho_p z}{144} \qquad (33.5)$$

and we have used the assumption that $g = g_c$. In this notation, including x, y, and z directional permeabilities and unit vectors $\hat{i}, \hat{j}, \hat{k}$, the Darcy velocities may be written as

$$\vec{v}_o = -\ddot{K} \cdot \lambda_o \nabla \Phi_o = -\lambda_o \left[\hat{i} K_x \frac{\partial \Phi_o}{\partial x} + \hat{j} K_y \frac{\partial \Phi_o}{\partial y} + \hat{k} K_z \frac{\partial \Phi_o}{\partial z} \right] \qquad (33.6)$$

$$\vec{v}_w = -\ddot{K} \cdot \lambda_w \nabla \Phi_w = -\lambda_w \left[\hat{i} K_x \frac{\partial \Phi_w}{\partial x} + \hat{j} K_y \frac{\partial \Phi_w}{\partial y} + \hat{k} K_z \frac{\partial \Phi_w}{\partial z} \right] \qquad (33.7)$$

and

$$\vec{v}_g = -\ddot{K} \cdot \lambda_g \nabla \Phi_g = -\lambda_g \left[\hat{i} K_x \frac{\partial \Phi_g}{\partial x} + \hat{j} K_y \frac{\partial \Phi_g}{\partial y} + \hat{k} K_z \frac{\partial \Phi_g}{\partial z} \right] \qquad (33.8)$$

We have used the dyadic notation $\ddot{K}$ to signify that permeability is a tensor of rank two. The expanded form of Eqs. (33.6) through (33.8) employs the common assumption that the coordinate axes of our reference system are aligned along the principal axes of $\ddot{K}$. As discussed in Chapters 7 and 8, and associated references, this assumption impacts the ability of the simulator to accurately model fluid flow.

Combining Eqs. (33.1) through (33.3) with Eqs. (33.6) through (33.8) gives

$$\nabla \cdot \frac{\ddot{K} \lambda_o}{B_o} \cdot \nabla \Phi_o - \frac{q_o}{\rho_{osc}} = \frac{\partial}{\partial t} \left(\frac{\phi S_o}{B_o} \right) \qquad (33.9)$$

$$\nabla \cdot \frac{\ddot{K}\lambda_w}{B_w} \cdot \nabla\Phi_w - \frac{q_w}{\rho_{wsc}} = \frac{\partial}{\partial t}\left(\frac{\phi S_w}{B_w}\right) \tag{33.10}$$

and

$$\nabla \cdot \ddot{K} \cdot \left[\frac{\lambda_g}{B_g}\nabla\Phi_g + \frac{R_{so}\lambda_o}{B_o}\nabla\Phi_o + \frac{\lambda_w R_{sw}}{B_w}\nabla\Phi_w\right] - \frac{q_g}{\rho_{gsc}}$$

$$= \frac{\partial}{\partial t}\left[\phi\left(\frac{S_g}{B_g} + R_{so}\frac{S_o}{B_o} + R_{sw}\frac{S_w}{B_w}\right)\right] \tag{33.11}$$

Equations (33.9) through (33.11) are equivalent to Peaceman's [1977] Eqs. (1-105) through (1-107) for a three-dimensional system, except we have also allowed gas to dissolve in the water phase. Our rate and coordinate system sign conventions also differ. If these differences are taken into consideration, the formulations are seen to be equivalent.

33.2 Introduction of the Capillary Pressure Concept

The presence of oil, water, and gas phase pressures in Eqs. (33.9) through (33.11) complicates the problem. We simplify the handling of the phase pressures and potentials in the flow equations by using the capillary pressure concept. Let us define the difference in phase pressures as

$$P_{cow} = P_o - P_w \tag{33.12}$$

and

$$P_{cgo} = P_g - P_o. \tag{33.13}$$

The differences P_{cow} and P_{cgo} are the capillary pressures for oil-water and gas-water systems, respectively. Experimentally P_{cow} and P_{cgo} have been observed to be principally functions of water and gas saturations, respectively. Using Eqs. (33.12) and (33.13) lets us write the water and gas phase potentials as

$$\Phi_w = P_o - P_{cow} - \frac{\rho_w z}{144} \tag{33.14}$$

and

$$\Phi_g = P_o + P_{cgo} - \frac{\rho_g z}{144} \qquad (33.15)$$

Combining Eqs. (33.9) through (33.11) with Eqs. (33.14) and (33.15) and rearranging yields

Oil

$$\nabla \cdot \ddot{K} \cdot \left(\frac{\lambda_o}{B_o} \right) \nabla P_o + CG_o - \frac{q_o}{\rho_{osc}} = \frac{\partial}{\partial t} \left(\phi \frac{S_o}{B_o} \right) \qquad (33.16)$$

Water

$$\nabla \cdot \ddot{K} \cdot \left(\frac{\lambda_w}{B_w} \right) \nabla P_o + CG_w - \frac{q_w}{\rho_{wsc}} = \frac{\partial}{\partial t} \left(\phi \frac{S_w}{B_w} \right) \qquad (33.17)$$

Gas

$$\nabla \cdot \left[\ddot{K} \cdot \left(\frac{\lambda_g}{B_g} + \frac{R_{so}\lambda_o}{B_o} + \frac{R_{sw}\lambda_w}{B_w} \right) \right] \nabla P_o + CG_g - \frac{q_g}{\rho_{gsw}}$$

$$= \frac{\partial}{\partial t} \left[\phi \left(\frac{S_g}{B_g} + \frac{R_{so}S_o}{B_o} + \frac{R_{sw}S_w}{B_w} \right) \right] \qquad (33.18)$$

The gravity and capillary contributions to the phase pressures have been collected in the terms CG_o, CG_w, and CG_g:

$$CG_o = -\nabla \cdot \ddot{K} \cdot \left(\frac{\lambda_o}{B_o} \right) \nabla \left(\frac{\rho_o z}{144} \right) \qquad (33.19)$$

$$CG_w = -\nabla \cdot \ddot{K} \cdot \left(\frac{\lambda_w}{B_w} \right) \nabla \left(\frac{\rho_w z}{144} + P_{cow} \right) \qquad (33.20)$$

and

$$CG_g = \nabla \cdot \ddot{K} \cdot \frac{\lambda_g}{B_g} \nabla \left(P_{cgo} - \frac{\rho_g z}{144} \right)$$

$$- \nabla \cdot \ddot{K} \cdot \left[\frac{R_{so} \lambda_o}{B_o} \nabla \left(\frac{\rho_o z}{144} \right) + \frac{R_{sw} \lambda_w}{B_w} \nabla \left(P_{cow} + \frac{\rho_w z}{144} \right) \right] \qquad (33.21)$$

Essentially our task is to solve Eqs. (33.16) through (33.18) and saturation constraint Eq. (32.20) for the four unknowns P_o, S_o, S_w, and S_g. All other physical properties in the equations are known, in principle, as functions of the four unknowns, or from field and laboratory data.

33.3 The Pressure Equation

The procedure used in BOAST4D to solve the flow equations requires that we first combine Eqs. (32.20) and (33.16) through (33.18) such that we have only one equation remaining for the unknown pressure P_o. We proceed by using the following shorthand for Eqs. (33.16) through (33.18):

Oil

$$L_o = \frac{\partial}{\partial t} \left(\phi \frac{S_o}{B_o} \right) \qquad (33.22)$$

Water

$$L_w = \frac{\partial}{\partial t} \left(\phi \frac{S_w}{B_w} \right) \qquad (33.23)$$

Gas

$$L_g = \frac{\partial}{\partial t} \left[\phi \left(\frac{S_g}{B_g} + \frac{R_{so} S_o}{B_o} + \frac{R_{sw} S_w}{B_w} \right) \right] \qquad (33.24)$$

where

$$L_o = \nabla \cdot \ddot{K} \cdot \frac{\lambda_o}{B_o} \nabla P_o + CG_o - \frac{q_o}{\rho_{osc}}, \qquad (33.25)$$

$$L_w = \nabla \cdot \ddot{K} \cdot \frac{\lambda_w}{B_w} \nabla P_o + CG_w - \frac{q_w}{\rho_{wsc}}, \tag{33.26}$$

and

$$L_g = \nabla \cdot \left[\ddot{K} \cdot \left(\frac{\lambda_g}{B_g} + \frac{R_{so}\lambda_o}{B_o} + \frac{R_{sw}\lambda_w}{B_w} \right) \right] \nabla P_o + CG_g - \frac{q_g}{\rho_{gsc}} \tag{33.27}$$

Recognizing that formation volume factors, gas solubilities, and porosity are functions of pressure, we use the chain rule to expand the accumulation terms (time derivatives) of Eqs. (33.22) through (33.24):

Oil

$$L_o = \frac{\phi}{B_o} \frac{\partial S_o}{\partial t} + \left[\frac{S_o}{B_o} \frac{\partial \phi}{\partial P_o} - \frac{S_o \phi}{B_o^2} \frac{\partial B_o}{\partial P_o} \right] \frac{\partial P_o}{\partial t}, \tag{33.28}$$

Water

$$L_w = \frac{\phi}{B_w} \frac{\partial S_w}{\partial t} + \left[\frac{S_w}{B_w} \frac{\partial \phi}{\partial P_o} - \frac{S_w \phi}{B_w^2} \frac{\partial B_w}{\partial P_o} \right] \frac{\partial P_o}{\partial t}, \tag{33.29}$$

Gas

$$
\begin{aligned}
L_g = {} & \frac{\phi}{B_g} \frac{\partial S_g}{\partial t} + \left[\frac{S_g}{B_g} \frac{\partial \phi}{\partial P_o} - \frac{S_g \phi}{B_g^2} \frac{\partial B_g}{\partial P_o} \right] \frac{\partial P_o}{\partial t} \\
& + \frac{\phi R_{so}}{B_o^2} \frac{\partial S_o}{\partial t} \\
& + \left[\frac{S_o R_{so}}{B_o} \frac{\partial \phi}{\partial P_o} + \frac{\phi S_o}{B_o} \frac{\partial R_{so}}{\partial P_o} - \frac{\phi S_o R_{so}}{B_o^2} \frac{\partial B_o}{\partial P_o} \right] \frac{\partial P_o}{\partial t} \\
& + \frac{\phi R_{sw}}{B_w} \frac{\partial S_w}{\partial t} \\
& + \left[\frac{S_w R_{sw}}{B_w} \frac{\partial \phi}{\partial P_o} + \frac{\phi S_w}{B_w} \frac{\partial R_{sw}}{\partial P_o} - \frac{\phi S_w R_{sw}}{B_w^2} \frac{\partial B_w}{\partial P_o} \right] \frac{\partial P_o}{\partial t}
\end{aligned} \tag{33.30}
$$

The saturation constraint

$$S_o + S_w + S_g = 1 \tag{33.31}$$

is now used to remove $\partial S_g/\partial t$ from Eq. (33.30). Differentiation of Eq. (33.31) by t and rearranging gives

$$\frac{\partial S_g}{\partial t} = -\frac{\partial S_o}{\partial t} - \frac{\partial S_w}{\partial t}. \tag{33.32}$$

Substituting Eq. (33.32) into Eq. (33.30) and simplifying yields

$$
\begin{aligned}
L_g =\ & \left(\frac{\phi R_{so}}{B_o} - \frac{\phi}{B_g} \right) \frac{\partial S_o}{\partial t} + \left(\frac{\phi R_{sw}}{B_w} - \frac{\phi}{B_g} \right) \frac{\partial S_w}{\partial t} \\[2mm]
& + \left\{ \frac{S_g}{B_g} \frac{\partial \phi}{\partial P_o} - \frac{S_g \phi}{B_g^2} \frac{\partial B_g}{\partial P_o} + \frac{S_o R_{so}}{B_o} \frac{\partial \phi}{\partial P_o} \right\} \frac{\partial P_o}{\partial t} \\[2mm]
& + \left\{ \frac{\phi S_o}{B_o} \frac{\partial R_{so}}{\partial P_o} - \frac{\phi S_o}{B_o^2} \frac{\partial B_o}{\partial P_o} \right\} \frac{\partial P_o}{\partial t} \\[2mm]
& + \left\{ \frac{S_w R_{sw}}{B_w} \frac{\partial \phi}{\partial P_o} + \frac{\phi S_w}{B_w} \frac{\partial R_{sw}}{\partial P_o} - \frac{\phi S_w R_{sw}}{B_w^2} \frac{\partial B_w}{\partial P_o} \right\} \frac{\partial P_o}{\partial t}.
\end{aligned}
\tag{33.33}
$$

Equations (33.28), (33.29), and (33.33) are three equations for the three unknowns P_o, S_o, S_w. Multiplying Eq. (33.28) by $(B_o - R_{so} B_g)$, Eq. (33.29) by $(B_w - R_{sw} B_g)$, Eq. (33.33) by B_g, and adding the results gives

$$\left(B_o - R_{so}B_g\right) L_o + \left(B_w - R_{sw}B_g\right) L_w + B_g L_g$$

$$= \left(B_o - R_{so}B_g\right) \frac{\phi}{B_o} \frac{\partial S_o}{\partial t} + \left(B_w - R_{sw}B_g\right) \frac{\phi}{B_w} \frac{\partial S_w}{\partial t}$$

$$+ B_g \left(\frac{\phi R_{so}}{B_o} - \frac{\phi}{B_g} \right) \frac{\partial S_o}{\partial t} + B_g \left(\frac{\phi R_{sw}}{B_w} - \frac{\phi}{B_g} \right) \frac{\partial S_w}{\partial t}$$

$$+ \left[B_o - R_{so}B_g \right] \left[\frac{S_o}{B_o} \frac{\partial \phi}{\partial P_o} - \frac{S_o \phi}{B_o^2} \frac{\partial B_o}{\partial P_o} \right] \frac{\partial P_o}{\partial t}$$

$$+ \left[B_w - R_{sw}B_g \right] \left[\frac{S_w}{B_w} \frac{\partial \phi}{\partial P_o} - \frac{S_w \phi}{B_w^2} \frac{\partial B_w}{\partial P_o} \right] \frac{\partial P_o}{\partial t} \qquad (33.34)$$

$$+ \left\{ S_g \frac{\partial \phi}{\partial P_o} - \frac{S_g \phi}{B_g} \frac{\partial B_g}{\partial P_o} + \frac{B_g S_o R_{so}}{B_o} \frac{\partial \phi}{\partial P_o} \right\} \frac{\partial P_o}{\partial t}$$

$$+ \left\{ \frac{B_g \phi S_o}{B_o} \frac{\partial R_{so}}{\partial P_o} - \frac{\phi B_g S_o R_{so}}{B_o^2} \frac{\partial B_o}{\partial P_o} \right\} \frac{\partial P_o}{\partial t}$$

$$+ \left\{ \frac{B_g S_w R_{sw}}{B_w} \frac{\partial \phi}{\partial P_o} + \frac{B_g \phi S_w}{B_w} \frac{\partial R_{sw}}{\partial P_o} - \frac{\phi B_g S_w R_{sw}}{B_w^2} \frac{\partial B_w}{\partial P_o} \right\} \frac{\partial P_o}{\partial t}$$

where some simplification has been performed. This mess can be greatly simplified by multiplying the bracketed terms and then combining with appropriate terms in the curly brackets. We also notice the terms involving time derivatives of S_o and S_w vanish identically. The result is

$$\left(B_o - R_{so}B_g\right) L_o + \left(B_w - R_{sw}B_g\right) L_w + B_g L_g$$

$$= \left[\left(S_g + S_w + S_o\right) \frac{\partial \phi}{\partial P_o} - \frac{\phi S_g}{B_g} \frac{\partial B_g}{\partial P_o} \right] \frac{\partial P_o}{\partial t} \qquad (33.35)$$

$$+ \left[\phi S_o \left(\frac{B_g}{B_o} \frac{\partial R_{so}}{\partial P_o} - \frac{1}{B_o} \frac{\partial B_o}{\partial P_o} \right) + \phi S_w \left(\frac{B_g}{B_w} \frac{\partial R_{sw}}{\partial P_o} - \frac{1}{B_w} \frac{\partial B_w}{\partial P_o} \right) \right] \frac{\partial P_o}{\partial t}$$

The oil, water, gas, rock, and total compressibilities are identified as

$$c_o = -\frac{1}{B_o}\frac{\partial B_o}{\partial P_o} + \frac{B_g}{B_o}\frac{\partial R_{so}}{\partial P_o},$$
(33.36)

$$c_w = -\frac{1}{B_w}\frac{\partial B_w}{\partial P_o} + \frac{B_g}{B_w}\frac{\partial R_{sw}}{\partial P_o},$$
(33.37)

$$c_r = \frac{1}{\phi}\frac{\partial \phi}{\partial P_o},$$
(33.38)

$$c_g = -\frac{1}{B_g}\frac{\partial B_g}{\partial P_o},$$
(33.39)

and

$$c_t = c_r + c_o S_o + c_w S_w + c_g S_g$$
(33.40)

respectively. Employing these definitions, Eqs. (33.25) through (33.27) and (33.31) in Eq. (33.35) gives

$$\left(B_o - R_{so}B_g\right)\left[\nabla \cdot \ddot{K}\cdot\frac{\lambda_o}{B_o}\nabla P_o + CG_o - \frac{q_o}{\rho_{osc}}\right]$$

$$+\left(B_w - R_{sw}B_g\right)\left[\nabla \cdot \ddot{K} \cdot \frac{\lambda_w}{B_w}\nabla P_o + CG_w - \frac{q_w}{\rho_{wsc}}\right]$$
(33.41)

$$+ B_g\left[\nabla \cdot \ddot{K} \cdot \left(\frac{\lambda_g}{B_g} + \frac{R_{so}\lambda_o}{B_o} + \frac{R_{sw}\lambda_w}{B_w}\right)\nabla P_o + CG_g - \frac{q_g}{\rho_{gsc}}\right]$$

$$= \phi c_t \frac{\partial P_o}{\partial t}.$$

Equation (33.41) is called the pressure equation because no explicit time derivatives of saturations are present. BOAST4D is coded to solve the three-dimensional, three-phase flow equations by first numerically solving the pressure equation for P_o, then using the results in Eqs. (33.22), (33.23) and (33.31) to find the phase saturations. This procedure is an example of a numerical method known as the IMplicit Pressure-Explicit Saturation (IMPES) procedure. Chapter 7 presents a discussion of alternative formulations for solving the fluid flow equations.

References

Fanchi, J.R., K.J. Harpole, and S.W. Bujnowski (1982): "BOAST: A Three-Dimensional, Three-Phase Black Oil Applied Simulation Tool", 2 Volumes, U.S. Department of Energy, Bartlesville Energy Technology Center, OK.

Sawyer, W.K. and J.C. Mercer (August 1978): "Applied Simulation Techniques for Energy Recovery," Department of Energy Report METC/RI-78/9, Morgantown, WV.

Cumulative References

Aguilera, R. (1980): **Naturally Fractured Reservoirs**, Tulsa, OK: PennWell Publishing.

Al-Hussainy, R. and N. Humphreys (1996): "Reservoir Management: Principles and Practices," *Journal of Petroleum Technology*, pp. 1129-1135.

Ammer, J.R. and A.C. Brummert (1991): "Miscible Applied Simulation Techniques for Energy Recovery – Version 2.0," U.S. Department of Energy Report DOE/BC-91/2/SP, Morgantown Energy Technology Center, WV.

Amyx, J.W., D.H. Bass, and R.L. Whiting (1960): **Petroleum Reservoir Engineering**, New York: McGraw-Hill.

Anderson, R.N. (1995): "Method Described for Using 4D Seismic to Track Reservoir Fluid Movement," *Oil & Gas Journal*, pp. 70-74, April 3.

Ausburn, B.E., A.K. Nath and T.R. Wittick (Nov. 1978): "Modern Seismic Methods – An Aid for the Petroleum Engineer," *Journal of Petroleum Technology*, pp. 1519-1530.

Aziz, K. (1993): "Reservoir Simulation Grids: Opportunities and Problems," *Journal of Petroleum Technology*, pp. 658-663.

Aziz, K. and A. Settari (1979): **Petroleum Reservoir Simulation**, New York: Elsevier.

Bear, J. (1972): **Dynamics of Fluids in Porous Media**, New York: Elsevier.

Beasley, C.J. (1996): "Seismic Advances Aid Reservoir Description," *Journal of Petroleum Technology*, pp. 29-30.

Benedict, M., G.B. Webb, and L.C. Rubin (1940): *Journal of Chemical Physics*, Volume 8, p. 334.

Blackwelder, B., L. Canales, and J. Dubose (1996): "New Technologies in Reservoir Characterization," *Journal of Petroleum Technology*, pp. 26-27.

Bradley, M.E. and A.R.O. Wood (Nov. 1994): "Forecasting Oil Field Economic Performances," *Journal of Petroleum Technology*, pp. 965-971 and references therein.

Brown, K.E. and J.F. Lea (Oct. 1985): "Nodal Systems Analysis of Oil and Gas Wells," *Journal of Petroleum Technology*, pp. 1751-1763.

Brock, J. (1986): **Applied Open-Hole Log Analysis**, Houston: Gulf Publishing.

de Buyl, M., T. Guidish, and F. Bell (1988): "Reservoir Description from Seismic Lithologic Parameter Estimation," *Journal of Petroleum Technology*, pp. 475-482.

Carr, N.L., R. Kobayashi, and D.B. Burrows (1954): "Viscosity of Hydrocarbon Gases Under Pressure," *Transactions of AIME*, Volume 201, pp. 264-272.

Carter, R.D. and G.W. Tracy (1960) "An Improved Method for Calculating Water Influx," *Transactions of AIME*, Volume 219, pp. 415-417.

Chin, W.C. (1993): **Modern Reservoir Flow and Well Transient Analysis**, Houston: Gulf Publishing.

Christie, M.A. (Nov. 1996): "Upscaling for Reservoir Simulation," *Journal of Petroleum Technology*, pp. 1004-1010.

Clark, N.J. (1969): **Elements of Petroleum Reservoirs**, Richardson, TX: Society of Petroleum Engineers.

Coats, K.H. (1969): "Use and Misuse of Reservoir Simulation Models," *Journal of Petroleum Technology*, pp. 183-190.

Collins, R.E. (1961): **Flow of Fluids Through Porous Materials**, Tulsa, OK: PennWell Publishing.

Craft, B.C. and M.F. Hawkins (1959): **Applied Petroleum Reservoir Engineering**, Englewood Cliffs, NJ: Prentice-Hall.

Craig, F.F. (1971): **The Reservoir Engineering Aspects of Waterflooding**, SPE Monograph Series, Richardson, TX: Society of Petroleum Engineers.

Crichlow, H.B. (1977): **Modern Reservoir Engineering – A Simulation Approach**, Englewood Cliffs, NJ: Prentice Hall.

Daccord, G., J. Nittmann and H.E. Stanley (1986): "Radial Viscous Fingers and Diffusion-Limited Aggregation: Fractal Dimension and Growth Sites," *Physical Review Letters* 56, 336-339.

Dahlberg, E.C. (1975): "Relative Effectiveness of Geologists and Computers in Mapping Potential Hydrocarbon Exploration Targets," *Mathematical Geology*, Volume 7, pp. 373-394.

Dake, L.P. (1978): **Fundamentals of Reservoir Engineering**, Amsterdam: Elsevier.

Dietrich, J.K. and P.L. Bondor (1976): "Three-Phase Oil Relative Permeability Models," SPE Paper 6044, *Proceedings of 51st Fall Technical Conference and Exhibition of Society of Petroleum Engineers and AIME*, New Orleans, LA, Oct. 3-6.

Dranchuk, P.M., R.A. Purvis, and D.B. Robinson (1974): "Computer Calculation of Natural Gas Compressibility Factors Using the Standing and Katz Correlations," *Institute of Petroleum Technology*, IP-74-008.

Earlougher, R.C., Jr. (1977): **Advances in Well Test Analysis**, SPE Monograph Series, Richardson, TX: Society of Petroleum Engineers.

Ebanks, W.J., Jr. (1987): "Flow Unit Concept – Integrated Approach to Reservoir Description for Engineering Projects," paper presented at the AAPG Annual Meeting, Los Angeles.

Englund, E. and A. Sparks (1991): "Geo-EAS 1.2.1 User's Guide," *EPA Report #600/8-91/008 Environmental Protection Agency-EMSL*, Las Vegas, NV.

Evans, W.S. (1996): "Technologies for Multidisciplinary Reservoir Characterization," *Journal of Petroleum Technology*, pp. 24-25.

van Everdingen, A.F. and W. Hurst (1949): "The Application of the Laplace Transformation to Flow Problems in Reservoirs," *Transactions of AIME*, Volume 186, pp. 305-324.

Fanchi, J.R. (1983): "Multidimensional Numerical Dispersion," *Society of Petroleum Engineering Journal*, pp. 143-151.

Fanchi, J.R. (June 1985): "Analytical Representation of the van Everdingen-Hurst Aquifer Influence Functions for Reservoir Simulation," *Society of Petroleum Engineering Journal*, pp. 405-406.

Fanchi, J.R. (1986): "BOAST-DRC: Black Oil and Condensate Reservoir Simulation on an IBM-PC," Paper SPE 15297, *Proceedings from Symposium on Petroleum Industry Applications of Microcomputers of SPE,* Silver Creek, CO, June 18-20.

Fanchi, J.R. (1990): "Calculation of Parachors for Compositional Simulation: An Update," *Society of Petroleum Engineers Reservoir Engineering*, pp. 433-436.

Fanchi, J.R. (1997): **Math Refresher for Scientists and Engineers**, New York: J. Wiley and Sons.

Fanchi, J.R. and R.L. Christiansen (1989): "Applicability of Fractals to the Description of Viscous Fingering," Paper SPE 19782, *Proceedings of 64th Annual Technology Conference And Exhibit of Society of Petroleum Engineers*, San Antonio, TX, Oct. 8-11.

Fanchi, J.R., K.J. Harpole, and S.W. Bujnowski (1982): "BOAST: A Three-Dimensional, Three-Phase Black Oil Applied Simulation Tool", 2 Volumes, U.S. Department of Energy, Bartlesville Energy Technology Center, OK.

Fanchi, J.R., J.E. Kennedy, and D.L. Dauben (1987): "BOAST II: A Three-Dimensional, Three-Phase Black Oil Applied Simulation Tool," U.S. Department of Energy, Bartlesville Energy Technology Center, OK.

Fanchi, J.R., H.Z. Meng, R.P. Stoltz, and M.W. Owen (1996): "Nash Reservoir Management Study with Stochastic Images: A Case Study," *Society of Petroleum Engineers Formation Evaluation*, pp. 155-161.

Fayers, F.J. and T.A. Hewett (1992): "A Review of Current Trends in Petroleum Reservoir Description and Assessing the Impacts on Oil Recovery," *Proceedings of Ninth International Conference On Computational Methods in Water Resources*, June 9-11.

Felder, R.D. (1994): "Advances in Openhole Well Logging," *Journal of Petroleum Technology*, pp. 693-701.

Gardner, H.F., L.W. Gardner, and A.H. Gregory (1974): "Formation Velocity and Density – the Diagnostic Basis for Stratigraphic Traps," *Geophysics* Volume 39, pp. 770-780.

Gassman, F. (1951): "Elastic Waves Through a Packing of Spheres," *Geophysics*, Volume 16, pp. 673-685.

Golf-Racht, T.D. van (1982): **Fundamentals of Fractured Reservoir Engineering**, New York: Elsevier.

Govier, G.W., Editor (1978): **Theory and Practice of the Testing of Gas Wells**, Calgary: Energy Resources Conservation Board.

Grace, J.D., R.H. Caldwell, and D.I. Heather (Sept. 1993): "Comparative Reserves Definitions: USA, Europe, and the Former Soviet Union," *Journal of Petroleum Technology*, pp. 866-872.

Haldorsen, H.H. and E. Damselth (1993): "Challenges in Reservoir Characterization," *American Association of Petroleum Geologists Bulletin*, Volume 77, No. 4, pp. 541-551.

Haldorsen, H.H. and E. Damsleth (April 1990): "Stochastic Modeling," *Journal of Petroleum Technology*, pp. 404-412.

Haldorsen, H.H. and L.W. Lake (1989): "A New Approach to Shale Management in Field-Scale Models," **Reservoir Characterization-2**, SPE Reprint Series #27, Richardson, TX: Society of Petroleum Engineers.

Harpole, K.J. (1985): **Reservoir Environments and Their Characterization**, Boston: International Human Resources Development Corporation.

Harris, D.G. (May 1975): "The Role of Geology in Reservoir Simulation Studies," *Journal of Petroleum Technology*, pp. 625-632.

He, W., R.N. Anderson, L. Xu, A. Boulanger, B. Meadow, and R. Neal (May 20, 1996): "4D Seismic Monitoring Grows as Production Tool," *Oil & Gas Journal*, pp. 41-46.

Hebert, H., A.T. Bourgoyne, Jr., and J. Tyler (May 1993): "BOAST II for the IBM 3090 and RISC 6000", U.S. Department of Energy Report DOE/ID/12842-2, Bartlesville Energy Technology Center, OK.

Hegre, T.M., V. Dalen, and A. Henriquez (1986): "Generalized Transmissibilities for Distorted Grids in Reservoir Simulation," Paper SPE 15622, *Proceedings of 61st Annual SPE Technical Conference and Exhibition*, Richardson, TX: Society of Petroleum Engineers.

Honarpour, M., L.F. Koederitz, and A.H. Harvey (1982): "Empirical Equations for Estimating Two-Phase Relative Permeability in Consolidated Rock," *Journal of Petroleum Technology*, pp. 2905-2908.

Isaaks, E.H. and R.M. Srivastava (1989): **Applied Geostatistics**, New York: Oxford University Press.

Jefferys, W.H. and J.O. Berger (1992): "Ockham's Razor and Bayesian Analysis," *American Scientist*, Volume 80, pp. 64-72.

Johnston, D.H. (1997): "A Tutorial on Time-Lapse Seismic Reservoir Monitoring," *Journal of Petroleum Technology*, pp. 473-475.

Joshi, S.D. (1991): **Horizontal Well Technology**, Tulsa, OK: PennWell Publishing Company.

Kamal, M.M., D.G. Freyder, and M.A. Murray (1995): "Use of Transient Testing in Reservoir Management," *Journal of Petroleum Technology*, pp. 992-999.

Kasap, E. and L.W. Lake (June 1990): "Calculating the Effective Permeability Tensor of a Gridblock," *Society of Petroleum Engineers Formation Evaluation*, pp. 192-200.

Kazemi, H. (Oct. 1996): "Future of Reservoir Simulation", *Society of Petroleum Engineers Computer Applications*, pp. 120-121.

Killough, J.E. (1995): "Ninth SPE Comparative Solution Project: A Reexamination of Black-Oil Simulation," Paper SPE 29110, *Proceedings of 13th Society of Petroleum Engineers Symposium on Reservoir Simulation*, Feb. 12-15.

Koederitz, L.F., A.H. Harvey, and M. Honarpour (1989): **Introduction to Petroleum Reservoir Analysis**, Houston: Gulf Publishing.

Lake, L.W. (April 1988): "The Origins of Anisotropy," *Journal of Petroleum Technology*, pp. 395-396.

Lantz, R.B. (1971): "Quantitative Evaluation of Numerical Diffusion," *Society of Petroleum Engineering Journal*, pp. 315-320.

Lee, S.H., L.J. Durlofsky, M.F. Lough, and W.H. Chen (1997): "Finite Difference Simulation of Geologically Complex Reservoirs with Tensor Permeabilities," Paper SPE 38002, *Proceedings of 1997 SPE Reservoir Simulation Symposium*, Richardson, TX: Society of Petroleum Engineers.

Lieber, Bob (Mar/Apr 1996): "Geostatistics: The Next Step in Reservoir Modeling," *Petro Systems World*, pp. 28-29.

Lough, M.F., S.H. Lee, and J. Kamath (Nov. 1996): "Gridblock Effective-Permeability Calculation for Simulation of Naturally Fractured Reservoirs," *Journal of Petroleum Technology*, pp. 1033-1034.

Louisiana State University (1997): " 'BOAST 3' A Modified Version of BOAST II with Post Processors B3PLOT2 and COLORGRID," Version 1.50, U.S. Department of Energy Report DOE/BC/14831-18, Bartlesville Energy Technology Center, OK.

Lynch, M.C. (1996): "The Mirage of Higher Petroleum Prices," *Journal of Petroleum Technology*, pp. 169-170.

Maddox, R.B. (1988): **Team Building: An Exercise in Leadership**, Crisp Publications, Inc.

Mattax, C.C. and R.L. Dalton (1990): **Reservoir Simulation**, SPE Monograph #13, Richardson, TX: Society of Petroleum Engineers.

Matthews, C.S. and D.G. Russell (1967): **The Reservoir Engineering Aspects of Waterflooding**, SPE Monograph Series, Richardson, TX: Society of Petroleum Engineers.

McCain, W.D., Jr. (1973): **The Properties of Petroleum Fluids**, Tulsa, OK: Petroleum Publishing.

McCain, W.D., Jr. (1991): "Reservoir-Fluid Property Correlations – State of the Art," *Society of Petroleum Engineers Reservoir Engineering*, pp. 266-272.

McIntosh, I., H. Salzew, and C. Christensen (1991): "The Challenge of Teamwork," Paper CIM/AOSTRA 91-19, *Proceedings of CIM/AOSTRA 1991 Technical Conference*, Banff, Canada.

McQuillin, R., M. Bacon, and W. Barclay (1984): **An Introduction to Seismic Interpretation,** Houston: Gulf Publishing.

Mian, M.A. (1992): **Petroleum Engineering Handbook for the Practicing Engineer**, Volumes I and II, Tulsa, OK: PennWell Publishing.

Millheim, K.M. (1997): "Fields of Vision," *Journal of Petroleum Technology*, p. 684.

Moses, P.L. (July 1986): "Engineering Applications of Phase Behavior of Crude Oil and Condensate Systems," *Journal of Petroleum Technology*, pp. 715-723; and F.H. Poettmann and R.S. Thompson (1986): "Discussion of Engineering Applications of Phase Behavior of Crude Oil and Condensate Systems," *Journal of Petroleum Technology*, pp. 1263-1264.

Murtha, J.A. (1997): "Monte Carlo Simulation: Its Status and Future," *Journal of Petroleum Technology*, pp. 361-373.

Nemchenko, N.N., M. Ya. Zykin, A.A. Arbatov, V.I. Poroskun, and I.S. Gutman (1995): "Distinctions in the Oil and Gas Reserves and Resources Classifications Assumed in Russia and USA – Source of Distinctions," *Energy Exploration and Exploitation*, Volume 13, #6, Essex, United Kingdom: Multi-Science Publishing Company.

Nolen, J.S. and D.W. Berry (1973): "Test of the Stability and Time-Step Sensitivity of Semi-Implicit Reservoir Simulation Techniques," **Numerical Simulation**, SPE Reprint Series #11, Richardson, TX: Society of Petroleum Engineers.

Norton, R. (Nov. 1994): "Economics for Managers," *Fortune*, p. 3.

Odeh, A.S. (1981): "Comparison of Solutions to a Three-Dimensional Black-Oil Reservoir Simulation Problem," *Journal of Petroleum Technology*, pp. 13-25.

Odeh, A.S. (Feb. 1985): "The Proper Interpretation of Field-Determined Buildup Pressure and Skin Values for Simulator Use," *Society of Petroleum Engineering Journal*, pp. 125-131.

Oreskes, N., K. Shrader-Frechette, and K. Belitz (1994): "Verification, Validation, and Confirmation of Numerical Models in the Earth Sciences", *Science*, pp. 641-646, Feb. 4.

Pannatier, Y. (1996): **VARIOWIN: Software for Spatial Data Analysis in 2D**, New York: Springer-Verlag.

Paterson, L. (January 1985): "Fingering with Miscible Fluids in a Hele-Shaw Cell," *Physics of Fluids,* Volume 28 (1), 26-30.

Peaceman, D.W. (1977): **Fundamentals of Numerical Reservoir Simulation**, New York: Elsevier.

Peaceman, D.W. (June 1978): "Interpretation of Well-Block Pressures in Numerical Reservoir Simulation," *Society of Petroleum Engineering Journal*, pp. 183-194.

Peaceman, D.W. (June 1983): "Interpretation of Well-Block Pressures in Numerical Reservoir Simulation with Nonsquare Grid Blocks and Anisotropic Permeability," *Society of Petroleum Engineering Journal*, pp. 531-543.

Pedersen, K.S., A. Fredenslund, and P. Thomassen (1989): **Properties of Oil and Natural Gases**, Houston: Gulf Publishing.

Prats, M. (1982): **Thermal Recovery**, SPE Monograph Series, Richardson TX: Society of Petroleum Engineers.

Raleigh, M. (Sept. 1991): *ECLIPSE Newsletter*, Houston: Schlumberger GeoQuest.

Raza, S.H. (1992): "Data Acquisition and Analysis for Efficient Reservoir Management," *Journal of Petroleum Technology*, pp. 466-468.

Reiss, L.H. (1980): **The Reservoir Engineering Aspects of Fractured Reservoirs**, Houston: Gulf Publishing.

Richardson, J.G., J.B. Sangree, and R.M. Sneider (1987a): "Applications of Geophysics to Geologic Models and to Reservoir Descriptions," *Journal of Petroleum Technology*, pp. 753-755.

Richardson, J.G., J.B. Sangree and R.M. Sneider (1987b): "Introduction to Geologic Models," *Journal of Petroleum Technology* (first of series), pp. 401-403.

Richardson, J.G. (Feb. 1989): "Appraisal and development of reservoirs," *Geophysics: The Leading Edge of Exploration*, pp. 42 ff.

Rosenberg, D. U. von (1977): **Methods for the Numerical Solution of Partial Differential Equations**, Tulsa, OK: Farrar and Associates.

Rossini, C., F. Brega, L. Piro, M. Rovellini, and G. Spotti (Nov. 1994): "Combined Geostatistical and Dynamic Simulations for Developing a Reservoir Management Strategy: A Case History," *Journal of Petroleum Technology*, pp. 979-985.

Ruijtenberg, P.A., R. Buchanan, and P. Marke (1990): "Three-Dimensional Data Improve Reservoir Mapping," *Journal of Petroleum Technology*, pp. 22-25, 59-61.

Sabet, M.A. (1991): **Well Test Analysis**, Houston: Gulf Publishing.
Saleri, N.G. (1993): "Reservoir Performance Forecasting: Acceleration by Parallel Planning," *Journal of Petroleum Technology*, pp. 652-657.
Saleri, N.G., R.M. Toronyi, and D.E. Snyder (1992): "Data and Data Hierarchy," *Journal of Petroleum Technology*, pp. 1286-1293.
Satter, A., J.E. Varnon, and M.T. Hoang (1994): "Integrated Reservoir Management," *Journal of Petroleum Technology*, pp. 1057-1064.
Sawyer, W.K. and J.C. Mercer (August 1978): "Applied Simulation Techniques for Energy Recovery," Department of Energy Report METC/RI-78/9, Morgantown, WV.
Schneider, F.N. (May 1987): "Three Procedures Enhance Relative Permeability Data," *Oil & Gas Journal*, pp. 45-51.
Schön, J.H. (1996): **Physical Properties of Rocks: Fundamentals and Principles of Petrophysics**, Volume 18, New York: Elsevier.
Sears, M. (June 1994): organizational development specialist at Bell Atlantic, quoted in *Journal of Petroleum Technology*, pp. 505.
Sheriff, R.E. (1989): **Geophysical Methods**, Englewood Cliffs, NJ: Prentice-Hall.
Slatt, R.M. and G.L. Hopkins (1990): "Scaling Geologic Reservoir Description to Engineering Needs," *Journal of Petroleum Technology*, pp. 202-210.
Spillette, A.G, J.G. Hillestad, and H.L. Stone (1986): "A High-Stability Sequential Solution Approach to Reservoir Simulation," **Numerical Simulation II**, SPE Reprint Series #20, Richardson, TX: Society of Petroleum Engineers.
Staff-JPT (Aug. 1994): "New Management Structures: Flat and Lean, Not Mean," *Journal of Petroleum Technology*, pp. 647-648.
Staff-JPT (May 1997): "SPE/WPC Reserves Definitions Approved," *Journal of Petroleum Technology*, pp. 527-528.
Standing, M.B. and D.L. Katz (1942): "Density of Natural Gases," *Transactions of AIME*, Volume 146, 140.
Stone, H.L. (Oct.-Dec. 1973); "Estimation of Three-Phase Relative Permeability and Residual Oil Data," *Journal of Canadian Petroleum Technology*, pp. 53ff.

Taggart, I.J., E. Soedarmo, and L. Paterson (1995): "Limitations in the Use of Pseudofunctions for Up-Scaling Reservoir Simulation Models," Paper SPE 29126, *Proceedings 13th Society of Petroleum Engineers Symposium on Reservoir Simulation*, San Antonio, TX, Feb. 12-15.

Tearpock, D.J. and R.E. Bischke (1991): **Applied Subsurface Geological Mapping**, Englewood Cliffs, NJ: Prentice Hall.

Telford, W.M., L.P. Geldart, R.E. Sheriff, and D.A. Keys (1976): **Applied Geophysics**, Cambridge: Cambridge University Press.

Thakur, G.C. (1996): "What Is Reservoir Management?" *Journal of Petroleum Technology*, pp. 520-525.

Thomas, G.W. (1982): **Principles of Hydrocarbon Reservoir Simulation**, Boston: International Human Resources Development Corporation.

Todd, M.R., P.M. Odell, and G.J. Hiraski (Dec. 1972): "Methods for Increased Accuracy in Numerical Reservoir Simulators," *Society of Petroleum Engineering Journal*, pp. 515-530.

Toronyi, R.M. and N.G. Saleri (1988): "Engineering Control in Reservoir Simulation," SPE 17937, *Proceedings of 1988 Society of Petroleum Engineers Fall Conference*, Oct. 2-5.

Uland, M.J., S.W. Tinker, and D.H. Caldwell (1997): "3-D Reservoir Characterization for Improved Reservoir Management," paper presented at the 1997 Society of Petroleum Engineers' 10th Middle East Oil Show and Conference, Bahrain (March 15-18).

Wiggins, M.L. and R.A. Startzman (1990): "An Approach to Reservoir Management," Paper SPE 20747, *Proceedings of 65th Annual Society of Petroleum Engineers Fall Meeting*, New Orleans, LA.

Williamson, A.E. and J.E. Chappelear (June 1981): "Representing Wells in Numerical Reservoir Simulation: Part I – Theory," *Society of Petroleum Engineers Journal*, pp. 323-338; and "Part II – Implementation," *Society of Petroleum Engineers Journal*, pp. 339-344.

Young, L.C. (1984): "A Study of Spatial Approximations for Simulating Fluid Displacements in Petroleum Reservoirs," **Computer Methods in Applied Mechanics and Engineering**, New York: Elsevier, pp. 3-46.

INDEX